AF255486

AWS

The Most Complete Guide to Learn Step by Step

Amazon Web Service

By

Robert Campbell

aws

Table of Contents

INTRODUCTION TO AMAZON CLOUDWATCH 82

CONCLUSION 88

Introduction

AWS provides a lot of web services. We will discuss about them here. Cloud computing, in the simplest terms, means storing and processing data and services over the Internet, rather than the hard drive of your computer.

Your hard disk is what cloud computing isn't about. That's called local storage and computing when you store data on–or run programs from the hard drive. Anything you need is physically close to you, which means you have fast and easy access to your data (for that one device, or those on the local network). Operating off the hard drive is how the tech industry has been operating for a decade.

The future of computing technology lies in the cloud. Which means that if you're not adapting your company to suit the cloud model your company will be left behind in this world of modern technology.

Cloud computing is when organizations share a network of freely accessible servers. Servers are stored on the Internet, allowing companies to handle data "in the cloud" instead of on a local server. It is a virtual space in which devices on the network can access data from anywhere.

While cloud computing has only picked up big momentum over the last two decades or so, the concept has been around since the 1960s. John McCarthy, a renowned computer scientist, introduced the idea when he invented a technology that would allow computation to be marketed as a commodity such as electricity or water. He indicated that each subscriber would only have to pay for the capacity they actually used and that certain users would be able to sell services to other users.

Amazon Web Services is a robust cloud platform developed by Amazon's e-commerce giant. It offers software-as - a-Service (SaaS), platform-as - a-Service (PaaS), and infrastructure-as - a-Service (IaaS) services. Think about the history of the electricity supply to grasp the logic of AWS.

Initially, factories will build their own plants to fuel their own facilities. Over time, governments and private investors have developed large power plants that supply electricity to numerous towns, factories, and homes. Under this new model, the factories will pay even less per unit of power due to the economies of scale enjoyed by the massive power plants. AWS was designed and built on the basis of a similar logic.

By 2006, Amazon had established itself as the world's largest online retailer, a role it still holds. Seamlessly running such a huge operation required a large and sophisticated infrastructure. Its imbued Amazon with deep expertise in the management of large-scale network and server networks.

As a result, AWS was launched in 2006 as Amazon tried to make accessible to companies and individuals the technology infrastructure it had developed and the expertise it had gained. AWS was one of the first pay-as-you-go (PAYG) computing models that could scale performance, storage, and computing based on the evolving needs of the user.

Getting start

AWS has a huge menu of offered services that are divided up into different sections for the convenience of its customers. These can be found from the "Products" section which can be accessed from the main page of the website.

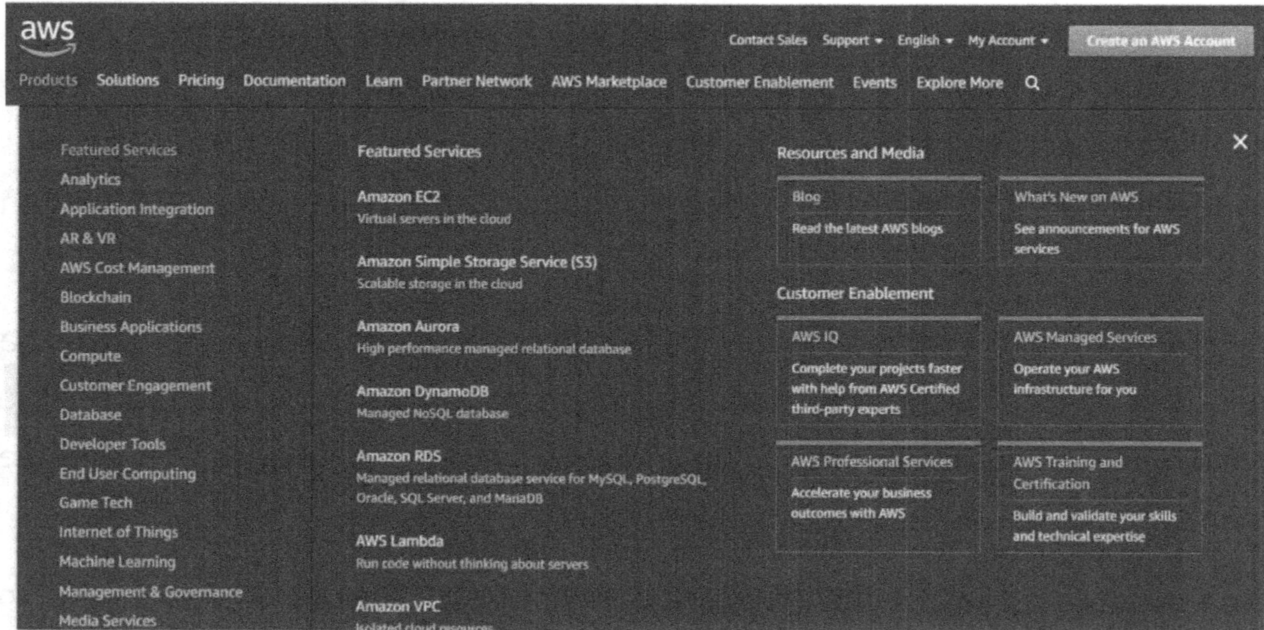

AWS has a massive infrastructure that spans the globe. You can see where their data centers are located by going to the "Global Infrastructure" page.

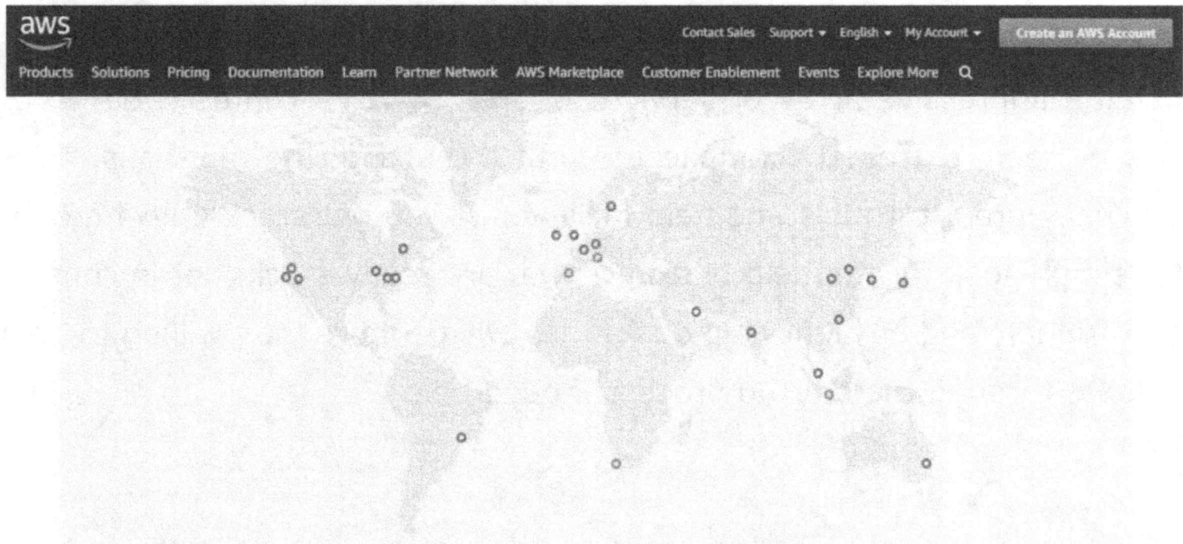

The orange dots on the page represent new upcoming regions, while the blue dots are current regions. This map is very useful in finding the best region and zone for your business. If you click on the blue dots, they will tell you the region and how many available zones there are in the region.

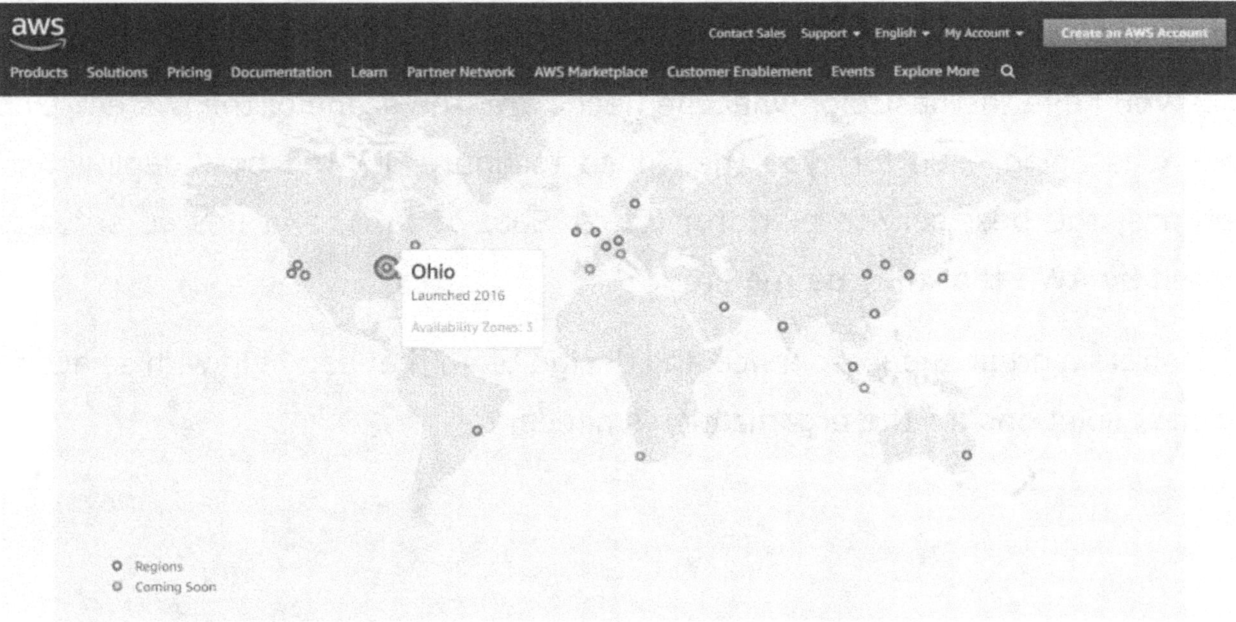

Where to Start?

Don't let the impressive array of services offered by AWS confuse you. AWS may look or seem extremely complicated, but the fact of the matter is, AWS also has really great tutorials and help facilities. If you get stuck anywhere on the site, simply look for the help button. If you are really stuck, get in contact with their friendly and efficient help desk; they will be more than willing to help you navigate through the startup process if need be.

Planning

The first step in the AWS journey is to scope out exactly what the cloud-based system requirements are.

To do this, spec out the system as if you are going to be installing an on-site networking system. Get all the details of what is required from the system, including the software and services that the system will need to support and run.

Once you have the basics of what the needs are, the scope of the system, and what it is being used for, you then need to figure out the best deployment methods, the best service platform (IaaS, SaaS, PaaS), and hosted services offered by AWS that may be required.

AWS offers a great support service and design team that can help with selecting the best solutions for the organization's needs.

The AWS Screen

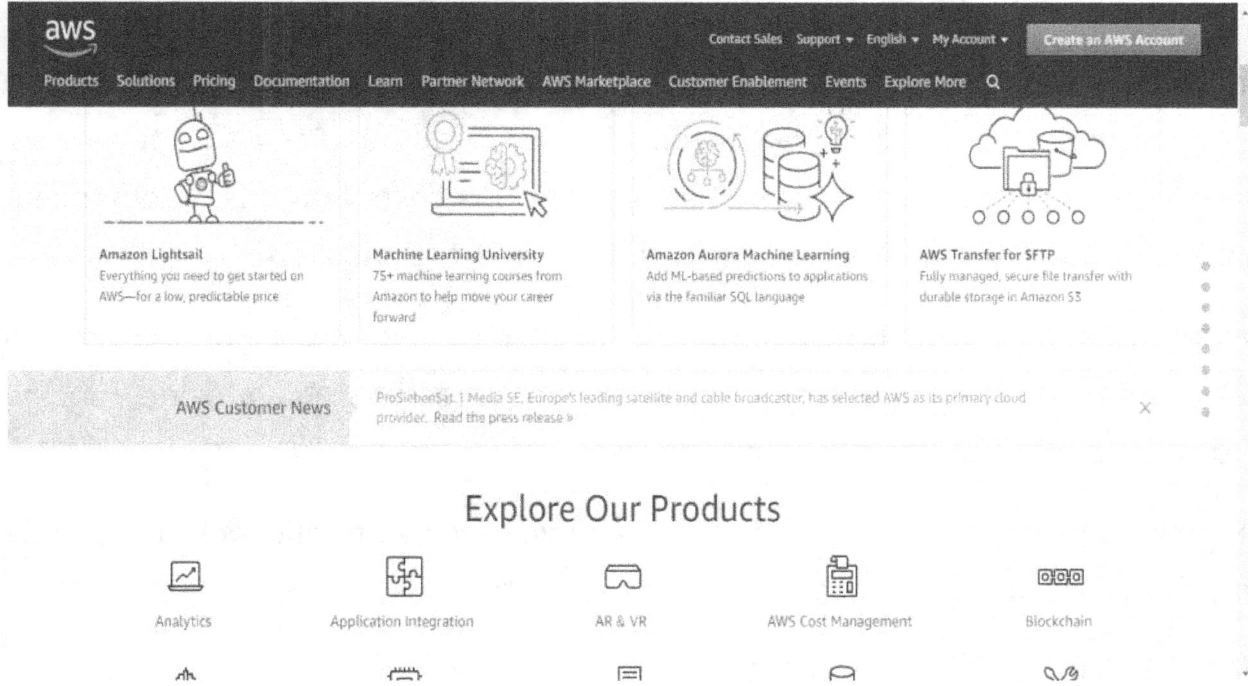

The AWS screen is quite a straightforward one. You can browse around at the services and various menu options listed on the site without having to register or login to the site.

Choosing the Correct Package or Service

AWS has an option to help with choosing the correct packages or services. This can be done by browsing through the products. These are broken down into groups pertaining to function. For instance, Analytics has a list of all the analytical services offered by AWS. Each of these services has a more in-depth overview to help the customer better understand their use.

They also have a Featured Services menu selection item which lists their most popular services. Another way to explore the services is to do so by going through the solutions section of the AWS website. The offered services are organized by means of what the hosted system is being used for, by industry, or by organization type.

Once you have a feel for the products and services offered, you will be able to better formulate the requirements for your organization. The next step would be to check out the pricing which lists the various costs and has a convenient section explaining how AWS pricing structure and options work.

Having all the information gathered, the next step is to create an account on AWS, enter all the necessary details, and choose the region, zone, pricing structure, deployment model, and services.

Once the services have been activated, you will receive notification and will be able to get started on the new cloud-based system.

What Are the Zones and Regions?

Amazon has very high-tech data centers that are located in different geographical areas. These geographical areas are the regions, and regions have a few separate locations called availability zones. Local zones are the areas where resources can be placed to make them closer to the end user.

How to Choose a Zone and Region?

Availability Zones allow an organization to distribute resources across them so there is no one point of failure in case one of the AWS zones goes down. If a company has instances that have been distributed across multiple zones, failover processes can be built in to make sure outages in one zone can be handled by another zone.

Local Zones allow for the distribution of services to be placed in a zone that is closer to the end user for their convenience. However, local zones are not available in every region. To check if there is a local zone in the required area you need to check on the website under "Geographic Locations."

Regions are where an instance of EC2 servers are hosted. Each region will be in a separate location to each other to ensure system resilience. With isolated instances of services located in various remote regions, AWS can offer a high level of service continuity.

Although regions can be specific to where the organization is located, regions can be selected for services to be closer to the end user or customer. This means if the client is located in South America, they can have instances located in North America, Europe, and so on.

Always check the pricing and service plan, as there are charges for the transferring of data between the regions.

Creating the Account

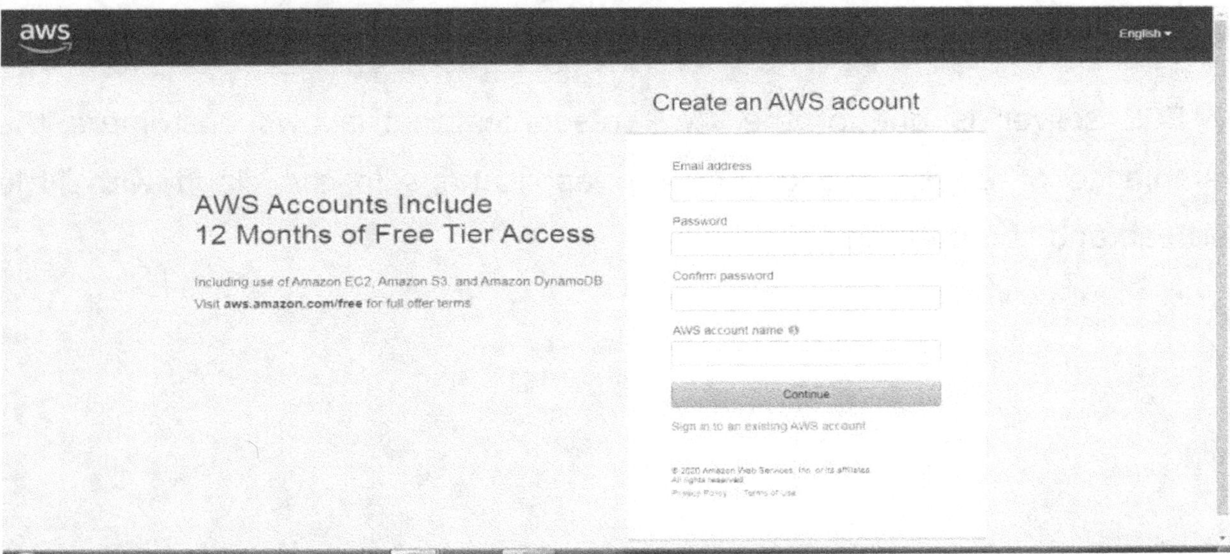

Create an account by selecting the "Sign Up" button on the top right-hand corner of the AWS web page. Enter the email address to be used for account transactions and correspondences. Set up a password and choose the AWS account name, click continue, and follow the instructions that will take you through the setup step by step.

At any time, if you are in need of assistance, AWS help, and support services are available 24/7.

AWS Management Console

AWS has a management console to help users manage and navigate through their services. It is a customizable graphical user interface (GUI) that allows the end user control over selected services. It helps the end user run services such as applications, storage, cloud infrastructure, and any other service they run over the AWS cloud.

EC2 Server

AWS offers many services that are quick and easy to launch at cost-effective prices. Nearly all of AWS packages are scalable and work on an elastic compute basis to ensure they are operating on only the resource amount they need.

The EC2 server is one of the AWS resources that allows customers the convenience of elastic capacity to run applications in the cloud with little overhead and commitment.

EC2: A Quick Overview

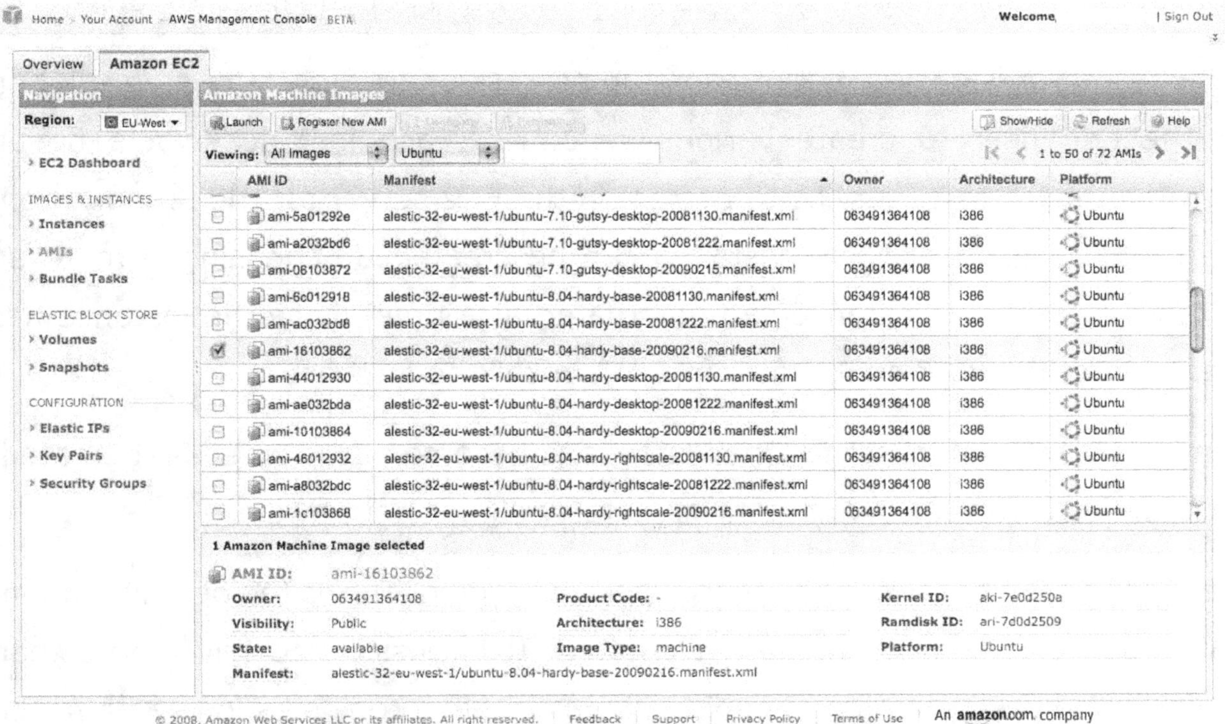

EC2 is the AWS elastic compute cloud that gives end users a secure, highly scalable cloud environment. It gives the customers complete control over their chosen resources and the security of operating within the AWS cloud.

EC2 allows for the establishment of virtual machines for different operating systems to run within the cloud. This increases a company's IT resource capacity, especially for big data and developers. Systems can be tested across multiple platforms without having to invest in expensive equipment to run virtual machines on.

EC2 helps its customers overcome previous compute power limitations they may have had. The fact that the AWS EC2 environment is scalable and elastic means that if a project requires more resources than at first was allocated, it is quick and easy to get more. There is no need for the red tape of change requests or procurement requests; all it takes is re-scaling the resource to meet the current demand. If that demand is no longer required, it is simple enough to scale back down and not pay for resources that are no longer necessary.

The EC2 server is easy to set up, and the process offers a step-by-step solution for the end user to create various instances. Each instance would be an operating platform or service to be run on the EC2 server for the organization. On the EC2 dashboard, the instance can be created by choosing "Create an Instance." Only one instance can be created at a time. Once the instance has been created to work on it, you simply "launch" the instance.

It may take a few minutes for the instance to become active when first created, but once it is, you can switch to it and start using it right away.

Users have a limit of twenty instances per EC2 region. This is the default amount that is set for any account that is created. It is possible to have more than twenty instances running on one EC2 server within a region as long as Amazon has approved it and made allowances for it.

AWS Free Tier & Pricing

The AWS pricing screen has a lot of helpful information on it. All the cloud hosting costs are completely transparent and kept updated. Before you select services, it is advisable to go to the AWS pricing screen. There is a section there that explains the different costs and pricing for various solutions. It also has a section to help an organization optimize its costs as well as a calculator to help calculate various costs.

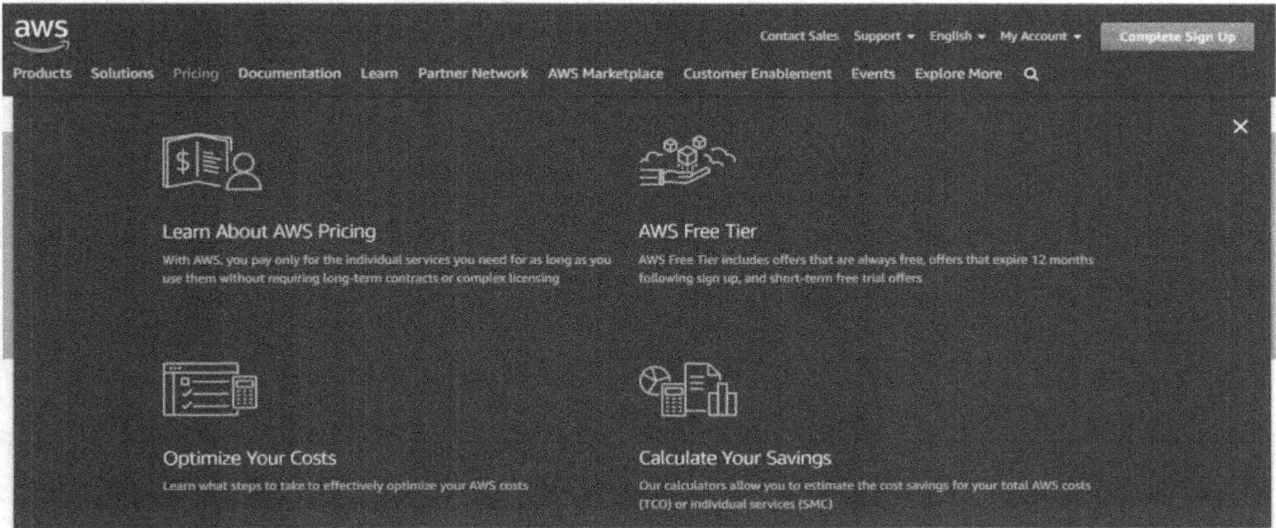

Free Tier

When first signing up with AWS, unless specified otherwise, the account is set to the free tier option. The free tier has a lot of the available resources and services, but it also has set limits that if exceeded will be billed for.

The free tier expires after a year, but if an organization requires more services than what the limits are for the free tier, these services will be charged to the customer. There are various one-time limits set on some of the services, and all other services that are exceeded get a pay-per-use rate charged. These pay-per-use charges will be the charges as explained in the AWS pricing information.

Small businesses, large enterprises, students, and organizations may all qualify to use the free tier. However, only one user account will be allowed to use the free tier account per organization. It should be noted that any services on the free tier that are used above the free tier limits are charged as a standard fee to the organization. This cost will be added to the total bill, which will include all the AWS accounts associated with an organization.

AWS offers three variations of its free tier. The first is an "always free" tier which has no expiration. The second variation is a short-term trial where end users can test out different services and solutions. And the third variation is the default 12-month free trial which an end user gets upon signing up with AWS.

Pricing

Each service or solution may have its own cost per usage, but AWS does have three standard pricing models.

Pay-as-you-go is the model that allows more flexibility in that you only pay for the service or resources that you need when you need them. For instance, a large project may require extra storage space for the duration of the project. This will be charged while the extra space is being used. When the project is finished and space is no longer required, the charge will be dropped as soon as space is no longer used.

Use more and pay less is a model that can help an organization cut costs for various services with a tiered pricing scheme that provides lower unit costs when you use more. This works similarly to contract pricing structures where buying a contract for two years instead of one saves you a few dollars a month. By opting for a larger storage capacity with S3, you get a tiered pricing scheme where the bigger the storage capacity costs less per unit than a small capacity, though the total cost will be higher.

Compute

Amazon's Cloud is officially known as Amazon Web Services or Abbreviated as "Amazon AWS ". On Amazon AWS there are many types of services Today I would like to introduce you to the first service which is named Amazon Elastic Compute Cloud or simply EC2

EC2's working principle is to create a Virtual Server that allows users to choose how much resources they need. For example, we can choose the size of the CPU, Memory and Disk as needed. At this point, some of you may be familiar or used to use VMware systems that have shared resources on the server into smaller VMs, which in the Amazon AWS system uses the same method.

However, the Amazon EC system 2 there will be no cost for server equipment and software license of the operating system such as the license of MS Windows Server. For using Amazon EC2 system, we will only pay according to the amount of resources we use. Have you seen that creating a computer system on Amazon EC2 is interesting, convenient, and low in cost?

Amazon Elastic Compute Cloud (EC2)

The EC2 is a cloud service that provides virtual servers (Amazon EC2 Instance), 2 types of data warehouses, as well as a load balancer (Load Balancer).

Adding an infinite number of disks with an almost infinite amount of storage, EBS (Known as Elastic Block Storage) is one of the kinds of storage in EC2. Its peculiarity is that the disks created using this technology are independent of the VPS nodes and are located on special Storage servers, in contrast to Instance storages, which are located directly on the virtualization servers.

Using EBS, you can add drives of any size to the running servers.

Creating a disk

Disk Management:

Elastic IP addresses make it possible to quickly change the server address, for example, in order to avoid DNS propagation - the time for updating a DNS zone around the world.

Creating instant images (Snapshot) allows you to create a cast of the disk and use it as a source for AMI (Amazon Machine Image), as well as for a simple OS backup.

Server types

EC2 servers can be described in the following table:

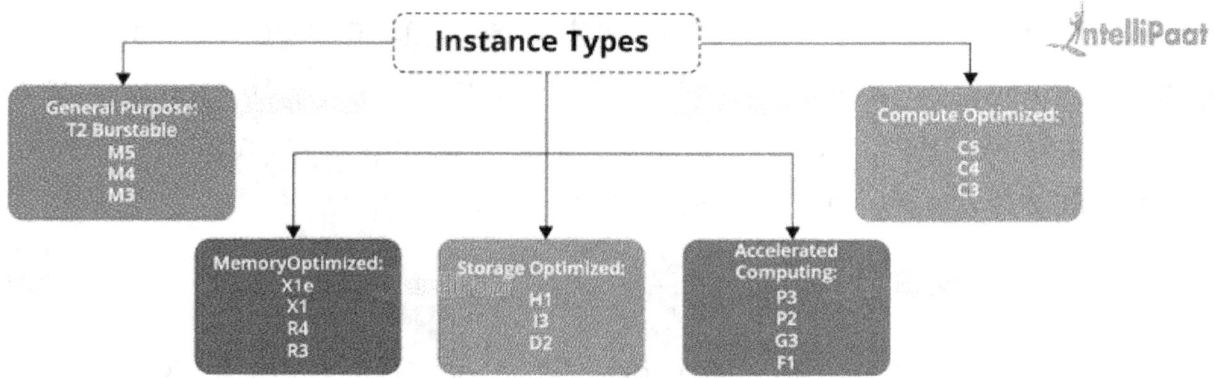

EC2 compute unit - a unit of measurement of processor performance, comparable with the performance of 1.0-1.2 GHz processors Opteron or Xeon.

Billing

EC2 is billed hourly; some subservices like EBS have monthly billing. For each of the subservice there is a separate billing at a pre-determined price per hour or per month.

Also, in EC2 instances there is a so-called reservation (Reservation) - it is paid immediately 3-4 months of the server's operation, after which, the server's operating hour is ~ 1.5 times cheaper. Reservations are convenient to use if EC2 is used on an ongoing basis - saving on face.

This time, let's look at the highlights of the Amazon EC2 system that facilitates users

1. It can create a VM to use quickly within minutes.

2. Users can choose the type of OS that they want, whether it is Windows or Linux.

3. Can choose the size of the hard disk that you want. Set up security system which can be easily customized

4. It can be extended to backup or Disaster Recover system easily and at a low cost.

Benefits of Amazon EC2

1. No hardware costs which saves a lot of investment costs.

2. It can build a VM, also known as a Virtual Machine, to be used within quickly. We can choose the type of OS (Windows, Linux) and the size of the resources that need a variety.

3. In case the system is growing or has more users. Can automatically increase the size and resources of the VM to be larger

Hybrid Cloud is another solution that is used in businesses or organizations that require a lot of data security, such as law firms that want to keep customer information private and secure. Or used in the financial and health business group that contains personal information of many users or customers

Hybrid hosting is designed for website owners or system administrators who want the highest level of data security while at the same time taking advantage of cloud service features that can be accessed from anytime anywhere and the flexibility of the system that can be expanded as needed

Hybrid Cloud is not a solution built for everyone, because it combines the advantages of private and public cloud services for the benefit of increased usage in response to applications that require security. High data security will work with data encryption or Encrypted technology that helps increase the security of using and accessing data. Including portability features that can be used with a variety of operating systems allowing more flexibility in use

Basic components

This hybrid cloud model has three basic infrastructure types: Public Cloud, Private Cloud, and Cloud Service / Management Platform.

As for the Public Cloud and Private Cloud, it is considered an independent component that allows you to store and protect data in the system with private services and at the same time be able to take advantage of the resources of Public Cloud for processing in order to get the best performance

Benefits

This hybrid cloud service has the ability to maintain privacy that can be accessed through a privacy infrastructure. This system will not send information via the public internet. But still the features of cloud services that can take advantage of public cloud resources

Hybrid hosting allows you to benefit from the flexibility of the system that can be expanded according to usage needs. And does not need to transfer all the data to the third-party data center you will receive support for working with the system when used within the organization. But whenever the workload exceeds the capacity that the Private System can support, it is also supported by the Public system, which helps to increase security and promote system stability.

Despite having many advantages But one of the limitations of the Hybrid Cloud is that the price is quite high and requires a rather complex system management Therefore, it is a solution that should be chosen when absolutely necessary and the organization has enough budget to support the cost of the system.

Storage

AWS storage

You should consider AWS Storage search for several reasons:

• Storage is an increasingly important problem for computing due to the recent and dramatic increase in the amount of data that companies use in their daily businesses. Even though traditional structured data (the database) is growing fairly quickly, the use of digital media (video) by businesses is exploding. IT companies use more and more storage and often use communications service providers (CSPs) such as Amazon to provide storage. The recent increase in Big Data, which refers to the analysis of very large data sets, is another factor in storage consumption. Companies are drowning in the data and many are struggling to manage their own on-site storage system.

- Storage is Amazon's first AWS offering. Therefore, storage is an important part of the AWS ecosystem, which includes highly innovative uses of its storage services by AWS customers over the years.
- Various AWS offerings depend on AWS storage, including Simple Storage Service (S3). Understanding AWS storage services helps you better understand how AWS offers depend on AWS storage.
- AWS continues to innovate and provide new storage services. Glacier, for example, brings yet another solution to a historical computing problem: archival archives.
- We will cover the four AWS storage services:
- Simple storage service (S3): Provides highly scalable object storage in the form of unstructured bit collections.

- Elastic Block Storage (EBS): Provides highly available and reliable volumes of data that can connect to a virtual machine, extract and connect to another virtual machine.
- Glacier: a data storage solution; Provides extremely robust and inexpensive file data storage and retrieval
- DynamoDB: storage of key values; provide highly scalable, high-performance storage based on tables indexed by data values called keys
- You might wonder why Amazon offers four different AWS storage services. This interesting question, which is at the heart of Amazon's unique cloud computing offering, aims to respond well to the flood of data.
- Scalability: Traditional methods simply cannot scale enough to handle the amount of data that companies now generate. The amount of data that companies need to manage exceeds the capacity of almost any storage solution.
- Fast: They cannot move data fast enough to meet the demands of enterprise storage solutions. To be honest, most business networks can't handle the level of traffic needed to skip all the pieces stored by businesses.
- Cost: Given the volume of data processed, established solutions are not economically viable, are not accessible to the size that businesses now need.

For these reasons, the storage problem is passed along to the local storage (for example, the hard disk on the server using the data). Over the past two decades, two other forms of traditional storage have entered the market: network storage (NAS) and network storage (SAN), which drag storage from the local server to the network where it is installed. When the server requests data, instead of searching for a local disk, it searches for it on the network.

The two types of network storage are very different (despite the similarity of their acronym). The NAS, which functions as an extension of the server's local file system, is used as a local file reads and writes work the same way as if the file were on the server. In other words, NAS gives the impression that it is part of the local server. SAN works very differently. Provides separate remote storage from the local server; This storage does not appear as local on the server. Instead, the server must use a special protocol to communicate with the SAN device. You can say that the SAN device provides separate storage that the server will need special arrangements to use. Both types of storage are still widely used, but much larger volumes of data do not allow NAS or SAN storage to support the requirements. As a result, new types of storage with improved features have emerged.

In particular, there are now two new types of storage available:

- Purpose: Reliably saves and retrieves unstructured digital objects
- Key value: manages structured data

In warehouses objects

Object storage is used to store objects, which are essentially digital bit collections. These pieces can represent a digital photo, an MRI scanner, a structured document such as an XML or video file of your cousin's embarrassing attempt at skateboarding in the steps of the public library (the one she presented at her wedding). Object storage provides reliable (and highly scalable) bit collection storage but does not require any bit structure. The structure is selected by the user, who must know, for example, whether an object is a photo (which can be modified) or an MRI scan (which requires a special application to view it). The user must know both the format and methods of manipulation of the object. The object storage service simply provides reliable storage of parts.

Object storage is different from file storage, which you can get to know better on a computer. File storage provides an upgrade function, different from object storage. For example, suppose you store the log log of a program. The program constantly adds new registry entries when events occur. create a new one

Using it every time an additional recording record is created would be incredibly annoying. On the other hand, the use of storage allows you to update the file permanently by adding new information. In other words, update the file when the program creates new records. Object storage does not offer updating. You can insert or retrieve an object, but you cannot change it. Instead, update the object in the local application and insert it into the object store. To save the new version the same name as the previous one, delete the original object before inserting the new object of the same name. The difference may seem minor, but it requires different approaches to managing the objects looked at.

Key-value distribution storage

Distributed key-value storage, unlike object storage, provides structured storage that is somewhat similar to a database, but significantly differs in providing additional scalability and performance. You can already use a relational database management system, a storage product commonly known as RDBMS. Your data lines have one or more keys (hence the key-value storage name) that support data manipulation. While RDBMS systems are incredibly useful, they are generally facing scale problems beyond a single server.

New, key-value distributed storage products are designed from scratch to support large amounts of data across multiple servers (perhaps thousands).

Key-value storage systems often use redundancy in hardware resources to avoid interruptions; This concept is important when using thousands of servers because they may experience hardware failures. Without redundancy, the entire storage system can be closed by a single server. The use of redundancy makes the key value system always available and, most importantly, your data is always available because it is protected from hardware failures. Dozens of key storage products are available. Many of them have been developed by leading web-based companies, such as Facebook and LinkedIn, to allow them to handle a considerable amount of traffic. These companies have reviewed and published the products under Open Source licenses. You (or someone else) can now use it in other environments.

Although key-value storage systems vary in different ways, they have the following common characteristics:

The data is structured with a unique key used to identify the registry in which all the remaining data resides. The password is almost always unique, such as a username, a unique username (for example, title_1795456) or a part number. This ensures that each disk has a unique key, which facilitates scale and performance.

Recovery is limited to the value of the key. For example, to search all records with a common address (where the address is not the key), each record must be examined.

Searching for multiple data sets with common data elements is not supported. RDBMS systems allow for combinations: for a given username in a data set, search all records in a second data set with the username in individual records. For example, to find all user-retrieved books from the library, join the user table (where the username is used to identify their library ID) and the payment table (where each book is listed with the username). the library ID of all those who have checked). You can use the combination function of an RDBMS system to ask this question; On the other hand, since key-value systems do not support combinations, the two tables must be mapped to the application level rather than the storage systems. Using this concept, commonly referred to as "application-based intelligence", union execution requires application "intelligence" and many additional codings.

Key-value storage represents a trade-off between usability and scalability, and trading is scalability and less usability.

This proliferation of storage types gives users a much richer set of options for managing the data associated with their applications. Although they gain a lot of flexibility and can adapt the storage solution to functional requirements, they also have a challenge: a broader set of skills is needed to manage more storage solutions. In addition, using a key-value solution requires hundreds or even thousands of servers. Fortunately, Amazon recognizes that all of these storage solutions are important despite the management challenges involved and offers four types of storage solutions. A user can choose the one that best suits his needs, without having to integrate in his application a solution that is not compatible with the necessary features. The need for storage flexibility explains why Amazon offers four types of storage. You may not need all four, many users only manage one or two. You need to understand all the AWS options it offers, so you can choose a new one instead of relying on the existing solution.

S3 storage basics

S3 objects are treated as web objects, that is, they are accessible to Internet protocols using a URL identifier.

Now, you may ask, what are the cubes and the key, given in the first example? A repository in AWS is a group of objects. The name of the tax is associated with an account. For example, the deposit called aws4dummies is associated with my aws4dummies account. The name of the compartment should not be the same as the name of the account; It can be something. However, the repository name is completely flat: each repository name must be unique among all AWS users. If you try to create a test bay name in your account, you will receive an error message because you can bet that your last earnings have been reached. (For your information, an account is limited to 100 compartments).

A key in AWS is the name of an object and serves as an identifier for locating the data associated with the key. In AWS, a key can be an object name or a more complex layout that requires a structure for organizing objects in a cube (such as in the cube / photos / catphotos / Cat + Photo). JPG, where / photos / catphotos is part of the object name). This convenient layout provides a family directory type or URL format for object names; However, it does not represent the true structure of the S3 storage system. It's simply a convenient and memorable method of calling objects, which allows users to easily track them. Although many tools present S3 storage as if they were in a familiar mapping organization (including the AWS management console), they do not mean how objects are stored in S3.

S3 Object Management

An S3 object is not a complex creature, it is just a byte collection. The service does not impose restrictions on the format of the object, it is up to you. The only limitation refers to the size of the object: an S3 object is limited to 5 TB (this is great).

AWS Database

Amazon Aurora

The Amazon Aurora solution is a database engine that is compatible with MySQL and PostgreSQL, and it combines the general speed and availability of many higher – end databases with the simplicity and economy of open source database solutions. The Aurora solution of Amazon can operate up to five times better than MySQL, but with the security and reliability of a high – end commercial database, all at a mere fraction of the normal cost.

Benefits

High Performance – Amazon Aurora can have up to five times the throughput of a standard MySQL database, and up to twice the throughput of PostgreSQL that runs on comparable hardware, performance standards which are generally matched by high – end commercial databases, but Aurora provides this at about only 10% of the commercial database solution cost. The largest Aurora instance can run up to five hundred thousand reads and one hundred thousand writes per second and read operations can be scaled up even further if the client wishes by using replicas that have a latency of only ten milliseconds.

Security – The Aurora solution can provide the client with multiple levels of security as needed. Among the tools available are the Amazon virtual private cloud, encryption at rest through the key management service of Amazon web services, and data encryption when in transit using the SSL. When data is contained within an instance of Aurora, the data in the storage is encrypted, as well as all the backups, replicas, and snapshots of said data.

MySQL and PostgreSQL Compatibility – The database engine of Aurora is entirely compatible with the MySQL 5.6 database using the InnoDB storage solution, meaning that any tools such as codes, drivers, applications that are already being used in existing MySQL databases can be migrated to Aurora with little to no modification. This also allows for migration of whole databases using the standard import and export methods, or through replication of the MySQL database's binlog.

Scalable – The Aurora database can be scaled anywhere from an instance running 2 virtual CPUs and 4 gigabytes of memory to one running 32 virtual CPUs and 244 gigabytes of memory. If further scaling is needed, up to fifteen low latency read replicas can be added, distributed across three availability zones. Furthermore, storage is scalable as well, with the Aurora database solution scaling from 10 gigabytes to 64 terabytes of storage as needed.

Availability and Durability - The Aurora solution offers a 99.99% availability guarantee. Recovery from any physical failure is transparent, and failover of instances uses less than thirty seconds. The storage used by Aurora is fault – tolerant and self – healing, and is replicated six times across multiple availability zones, and is also constantly redundantly backed up to the S3 solution of Amazon.

Fully Managed – The Aurora database is fully managed by the service, meaning that the client no longer needs to worry about mundane management tasks such as hardware provisions, software patches, monitoring, or backing up, as these tasks are all automatically done for the client. In addition, the database is automatically backed up to the S3 storage solution.

Amazon RDS

The Relational database Service, or the RDS, is a service provided by Amazon that makes it easy and convenient for clients to be able to create and scale a relational type database in the Amazon Web Services Cloud, providing economical and resizable capacity while automatically managing the mundane database administration tasks, which allows the clients to focus on other things and save time and energy. The RDS offers six choices in terms of database engines, from Amazon's own Aurora Database, to MariaDB, Oracle, SQL Server, MySQL, and PostgreSQL.

Benefits

Fast and Convenient – The RDS solution makes it much easier for a client to move from the conception of their project to full deployment. The client has the option of using Amazon Web Services' management console, the RDS CLI, or even API calls to be able to access the relational database within a few minutes, without the need for tedious infrastructure provisioning, or download, installation, and maintenance of database software.

Scalable - The RDS solution allows the user to scale the needed computational and storage resources using a few clicks of the mouse or a simple API call, needing little to no downtime. In fact, many RDS engine types allow the client to run multiple read replicas to offload traffic from the primary instance of their database, making it scale to more data access demands.

Available and Durable – The RDS solution runs on the Amazon infrastructure, which is highly reliable and offers a 99.99% availability guarantee. If the user needs to provision a multi availability zone database instance, the RDS solution immediately synchronously replicates data to a standby instance in another availability zone to ensure data redundancy. In addition, standard features such as automatic backups, snapshots, and host replacement are all contained by the Amazon RDS.

Secure - Amazon RDS makes it much easier for the client to control who has access to their database. The RDS allows the client to isolate any database instances they have by running it within the virtual private cloud, and the RDS also allows the client to connect any existing infrastructure they have to the RDS solution through a virtual private network. In addition, RDS engines also offer data encryption when at rest and in transit.

Inexpensive - Thanks to the scale of Amazon's services, the client pays a relatively low rate for the computing capacity that they actually make use of. There are multiple purchasing options available to the client, and the following are a few examples of payment options:

On Demand Instances – This option allows the client to pay for capacity on an hourly basis, which is useful if the client does not yet want to set up a long-term arrangement. This setup allows the client to scale their capacity depending on what their program needs, and they will only pay for the instances that they actually use. This setup is beneficial for the client that wishes to avoid having to plan for the costs of hardware maintenance.

Reserved Instances – Reserved instances can provide the client with a large discount as compared to on –demand instance pricing, up to a seventy – five percent discount, but this would require the client to plan and reserve the instances that they will use in advance, reducing their possible flexibility.

Amazon DynamoDB

The DynamoDB service of Amazon is a fast and highly flexible database service that runs NoSQL, designed for all applications that need single digit MS latency regardless of scale. This database is fully managed and supports document data models, as well as key – value models. The performance rates of DynamoDB as well as the flexible data model make it a highly recommended database solution for gaming, mobile, web applications, as well as for Internet of things applications.

Benefits

Rapid and Consistent Performance - The Amazon DynamoDB solution was specially created to deliver consistent, rapid performance at whatever scale for any and all applications. Average service-side latencies are normally in single-digit ms. As the data volumes grow and program performance demands begin to increase, the DynamoDB solution makes use of automatic partitioning and SSD technology to meet the throughput requirements and deliver low latencies no matter the scale.

Highly Scalable - When the client creates a table, all they have to do is designate the amount of request capacity they will need. If the client's requirements change, they can simply modify their request capacity by making use of the management console or the dynamoDB API, allowing the service to manage all scaling needs, while ensuring throughput levels are maintained even during the scaling process.

Fully Managed – The DynamoDB database is a fully managed database service, meaning that the client simply has to create a table and set their desired throughput, and the service will be responsible for all the other tasks. This frees up the client from having to worry about mundane tasks such as hardware provisioning, software patches, setting up and configuring servers, or maintaining and partitioning the database and its data instances.

Event – Driven Programming – The DynamoDB service has a high level of integration with the Lambda service of Amazon Web Services, allowing it to provide "triggers" that allow the client to manage applications that react to changes in data.

Fine – grained Access – The DynamoDB service also integrates with the Identity and Access management of Amazon in order to allow for a high degree of control for access by the client, allowing the client to assign unique credentials to each user, designating what exactly each user is allowed to access.

Flexibility – The Amazon DynamoDB service supports key – value data models as well as document data structures, allowing the client a high degree of flexibility to design the best architecture for their needs.

Migration

Amazon Glacier

Amazon Glacier is a protected, solid, and amazingly easy cloud storage class for information documenting and long-haul backups. This is intended to convey 99.9% toughness and give far-reaching security and consistency capacities that can assist meet with the most stringent administrative demands. Customers can store information for as low as $1 per terabyte every month, noteworthy reserve funds contrasted with on-premises arrangements. To minimize expenses for fluctuating recovery needs, Amazon Glacier gives three choices to access to documents, from a couple of moments to a few hours.

Let's discuss some of its benefits enlisting its durability and cost-effective approach for developers.

Recoveries as quick as 1-5 minutes: The Amazon Glacier storage class gives three recovery alternatives to accommodate your utilization case. Sped up recoveries regularly return information in 1-5 minutes and are incredible for Active Archive use cases. Standard recoveries commonly complete between 3-5 hours and function admirably for less time-delicate needs like reinforcement information, media altering, or long-haul examination. Mass recoveries are the most reduced cost recovery choice, returning a lot of information inside 5-12 hours.

Most comprehensive security and compliance capabilities: The Amazon Glacier and Glacier Deep Archive storage classes offer a refined mix with AWS Cloud Trail to log, screen and hold storage API call exercises for reviewing, and supports three unique types of encryption.

Unrivaled durability and scalability: The Amazon Glacier storage classes run on the world's biggest worldwide cloud infrastructure and were intended for 99.9 % of solidness. Information is naturally circulated over at least three physical Availability Zones that are geologically isolated inside an AWS Region.

Minimal effort: Amazon Glacier's document is intended to be the least cost Amazon storage class, enabling you to file a lot of information at extreme ease. This makes it practical to hold each information you need for use cases like information lakes, investigation, Internet of Things, AI, consistence, and media resource documenting.

Amazon Snowball

This is a petabyte-scale data transport system that you can use. This is one of the best for larger companies that have a lot of data to move, and it won't cost as much. You can move analytics, video libraries, image repositories, backups, to even shutdowns, and tape replacement sand application migration. It's simple, fast, and more secure than ever before. You can create a job in AWS management and the device will be shipped to you. Once this is complete, the data can be returned as well.

How AWS Migration is done?

Data migration tends to be an easy process as it means moving the data from one place to another in the layman's language. But, as the process involves different phases, it is more complicated. Let me now think about migration's different phases:

Phase 1: Discovery

We don't need to move your entire business to the cloud. That's where segregation is important. You have to identify applications that can and cannot be migrated. This is all about the first phase.

Phase 2: Assessment

AWS provides different ways to migrate the request, for instance, depending on the details. AWS snowball, AWS snowmobile, direct connection with AWS, etc. Once you've chosen the right way to move your data, look for the resources you'll need to do that as well. Let's now discuss the various ways in which AWS Cloud data is stored.

 Phase 3: Proof of Concept (POC)

We are migrating, you've got to figure out how and where you're going to store it. The whole reason for moving to AWS is to minimize costs. You will test your workload in this phase and understand AWS Storage Service, its benefits, limitations, and the necessary security checks.

Phase 4: Application Migration to AWS

Now that you have all the pre-requisites with your on-site data repositories, such as the blueprint, migration tools, assignment list, backups, and synchronization. Finally, you can move your project to AWS Cloud. Once you've moved your plan to cloud, the added benefits you get are consistency and longevity. Let's look at the improvements that AWS brings to your Phase 5 architecture.

Phase 5: Enterprise Cloud Operations

You have already moved to AWS at this stage, and AWS will bring changes to your current infrastructure that you will need to integrate. Therefore, you need to make sure that after the migration you have a 247-support team to keep track of system maintenance and upgrades. So, this was about and how to incorporate the various phases of AWS Migration. Let's explore the AWS migration strategies.

Application Migration Strategies' The 6 R's'

Depending on the architecture, the complexity of migrating existing applications varies, Amazon came up with different strategies that they commonly called 6 Rs. Let's look at each one of them:

Rehost - You've got your software ready, and you can just rehost it on AWS, this is termed as "Lift and Change." Using a third-party exporting tool, you remove your services and applications from your hosting environment and shift them to the cloud.

Replatform: You have an outdated version of your application running on your hosting system, so you need to update and then rehost your software. Replatform is a "Lift and Shift" change. It involves optimizing the architecture of the cloud to achieve the benefits without changing the application's core architecture.

Repurchase: Many systems would not be compatible with the new architecture. You need to buy new software for the new architecture in that situation. AWS Marketplace offers a wide range of products with a "Pay as you Use" model as well. Repurchase is also called "drop and shop" where you upgrade, ease implementation, accept the new architecture, and make changes to the existing model.

Refactor: You want to add new features, scale up existing business model, and performance limitations that are difficult with the existing environment. Although the solution is a bit expensive, you rethink your needs. In the longer term, we need optimizing the business by switching to a service-oriented architecture (SOA).

Retire: You can differentiate between valuable and useless tools after AWS Migration. Therefore, you are cutting off all assets that are no longer useful to the company and creating a plan around the new resources. This will reduce the additional cost. You can now focus on maintaining the assets used by the new business model with fewer things to worry about.

Retain: As you know, you have to migrate the parts of your project. You can just use any of the above techniques. And develop a plan to maintain certain applications that are not yet ready to move to the cloud or the applications that have been updated recently, according to your business model.

Networking and Content Delivery

In the AWS scheme of things, networking is a big deal. Without it, there would be no sending and receiving network traffic from any of the AWS instances.

In networking, however, as in all other pieces of AWS design, Amazon came up with solution that is clearly ingenious— but just as clearly different from the traditional solutions that most people are familiar with. And for the same reasons, it broke ground in other facets of its design: increasing size and promoting automation, AWS took the path less traveled.

We do so over a network as machines speak to each other. This talking activity takes place on a TCP / IP network for the purpose of computing done around the globe. The standard of the TCP / IP network utilizes the concept of layers to illustrate how communication takes place.

The layers are numbered 1, 2, and 3, in this model:

1. The physical layer (Layer 1): is associated with the cables that sit in your office, or how your wireless access point speaks to your computer's wireless card.

2. The data-link layer (Layer 2): controls the data flow between network entities (hosts, domain names, subnet, whatever) that reside on the same network; this local area network (LAN) is dedicated to a single organization.

Usually, these organizations have a network interface card (NIC), each of which has a unique identifier— their Media Access Control (MAC) address. Layer 2 sets out how two entities that have MAC addresses can send data to each other.

3. The network layer (Layer 3): controls the flow of data between network entities residing on different networks. (Note that this data is sent via NIC, a convenient piece of hardware that is stored on a server.) Users communicate across multiple LANS in this wide-area network (WAN) and can't count on being connected on the same local physical layer. Layer 3 most often works by using the Internet Protocol (IP) to communicate using a logical addressing scheme (so-called, logically enough, IP addresses). IP addresses most commonly have four digits — say, 10.1.2.3 — with eight-bit data sets representing each digit.

Digital LANS

Keeping data private in a virtual networking environment (and do not forget that this is precisely what a cloud computing service provides at its core), how can you guarantee one user that their data is not available to another user? Obviously, one way is to create separate physical networks and allow each user account to have its own local area network, but that would be a logistical (and extremely expensive) nightmare. In addition, this method would require each user to have their own router to communicate all of their Layer 3 traffic to other, outside users.

Routers have been upgraded to provide specific users with Virtual LANs (VLANs) that essentially cordon off sections of larger, shared networks. Within that VLAN, traffic flows through Layer 2; any traffic flows through Layer 3 to other parts of the shared network, or out over the Internet.

Most hosting companies use VLAN technology to assign each customer a VLAN so that their computers are segregated from the computers of other customers.

This strategy, which provides the customers with a secure networking solution, tells them that their network traffic is immune from interception.

In general, during account setup, most hosting companies do all the work related to manually assigning and configuring VLANs. A network administrator will access the provider's router and configure the new customer's VLAN. The computers of the customer are then placed on the newly configured VLAN, and network traffic flows over it to them.

As hosting companies have moved into cloud computing, this practice of creating a VLAN for each new customer has almost universally continued, with new virtual machines assigned to the VLAN's address space. Depending on the cloud infrastructure of the provider, this VLAN can be configured manually or automatically. Continued use of VLANs within these environments makes sense, especially since many providers offer both hosting and cloud computing from the same facility; using a consistent VLAN approach allows for resource sharing and infrastructure simplicity.

This cloud computing utilization of VLANs has some drawbacks:

1. A delay in setting up of the account: Cloud computing services who have to manually build and configure VLANs place a delay in setting up the initial account. Some customers find the pause inconvenient; others see using that cloud computing provider as a barrier.

2. A restriction on the number of VLANs a router can manage: While this constraint can be overcome through the use of multiple routers, it adds difficulty to the network of the provider.

3. A limit on the number of computers that can be attached to a specific VLAN: Although many customers are unaffected, this limit is an unacceptable problem for web-scale applications that may require hundreds (if not thousands) of computers.

The Amazon alternative to VLANs Because Amazon wants to avoid the scaling constraints of VLAN technology in its cloud service, the VLAN approach is obviously unacceptable, for these reasons:

1. Limiting the number of VLANs would limit the number of customers Amazon could support with its AWS service. When Amazon first sketched out its plans for AWS, hundreds of thousands of different customers were expected to use AWS eventually, so this limitation was too restrictive.

2. The limitation of a customer's number of computers within a single VLAN would limit the number of instances that might be used in its applications. Ama spanned hundreds, if not thousands, of cases, so it also wanted its customers to. A solution that restricts the number of computers each customer uses is clearly unacceptable.

Consequently, Amazon built the network very differently from traditional methods, and with these features, Amazon itself had experience with its application collection implementing a networking design: 3. Using Layer 3 technology across the infrastructure: all traffic is directly based on the IP address, with no reliance on MAC addressing from Layer 2.

4. The specifications that an IP address and all traffic to that instance be allocated to each instance must be driven by IP address: this is valid if traffic originates within AWS or externally — no exceptions.

5. No use or support of VLAN technology: Amazon has one or more IP address ranges within each area, and IP addresses are allocated randomly to customer instances within those address ranges. A corollary to this strategy is that all AWS IP addresses belong to Amazon, not to the client. So, if a customer decides to move their website from their own data center to AWS, they will have a new IP address at the website.

AWS Direct Connect

The fact that all network traffic between AWS and non-AWS resources travels over the public Internet poses a major problem: although Internet connectivity is offered by very large service providers who have invested a great deal of money in their networks, the bandwidth and latency levels available to end-users are highly variable and may be unacceptable.

At the birth of the Internet, the seeds for these kinds of problems existed. By its very nature, the Internet is a shared network, where millions of packets of computers are intermingled as they are sent over the network. Packets from your machine jostle with those of everybody else. The upside is that a shared network is much simpler (say Hello to email and Facebook); the drawback is that there is much less reliable efficiency and throughput in a shared network.

It's not a big deal for you and me. If a Netflix video runs slowly, it's not an earth-shattering issue, and many of the things we do aren't greatly affected by network problems. E-mail, for example, usually works the same, with network throughput varying by as much as 1,000 per cent.

However, inconsistent network throughput can be a big issue for companies.

Okay, when you can't watch a video, you're going about your business and doing something else. However, if an employee is unable to watch a safety video, it can affect her ability to work, and paying someone who cannot work is a big problem.

From the point of view of many businesses, another issue may occur: Internet traffic flows over a shared network and may require unauthorized access to data from a client. Sending traffic over a publicly accessible network is a no-no for certain businesses or certain types of data.

With Direct Connect, Amazon addresses the issue of traffic flowing through the public Internet: it allows a user to set up a private circuit between his data center and AWS to allow traffic to flow through a dedicated network connection without using the public Internet.

Direct Connect dedicated network connections, like Equinox, can be made from AWS to either a company's own data center or a public carrier. The company requesting the link to the Direct Connect network may have their servers located at the site of the public carrier or have a second connection to the company's own data center from the public carrier.

A dedicated network connection naturally addresses the issue of the protection of packets. A business using Direct Connect can be confident that its traffic on the network is free from prying eyes. In addition, to further ensure data security, it can enforce a virtual private network (VPN) between its AWS instances and its own data center. (Describing VPNs and how they work is beyond this book's scope, but it's enough to say they use clever software to encrypt data traveling across insecure networks— like the public Internet.) Amazon offers two levels of Direct Connect bandwidth: 1 Gbps and 10 Gbps.

For most connectivity needs, the former should be sufficient; the latter is sufficient for all but the most demanding high-performance computing and corresponds to the highest level of throughput within AWS itself.

Direct Connect comes with a financial arrangement similar to AWS: you only use Direct Connect when you need it, and you only pay for it when you use it.

For the 1 Gbps variant, Direct Connect costs $.30 per hour, and for the 10 Gbps variant $2.25 per hour. You don't pay for inbound network traffic as you might expect, and outbound traffic runs from $.03 to $.11 per gigabyte, depending on the region.

High-Performance AWS Networking

One concern about AWS networking is related to its performance — in my view, using the whole AWS infrastructure is the primary challenge. Amazon offers few specifics of its infrastructure, but I believe the company has 1 Gbps networking equipment in its data centers, which could provide technically acceptable performance for most applications, with 10 Gbps networking equipment used for more challenging AWS services, such as high-throughput instances. In other words, you're sharing the AWS network with some true hogs in the bandwidth, and the bandwidth competition can definitely affect the throughput of your application.

Most AWS users generally see approximately 100 Mbps throughput during their daily use of AWS in inter-instance network traffic. The problem is that although this average may be perfectly acceptable for many applications, the varying network load can alter that throughput significantly. Of course, 100 Mbps may be perfectly acceptable for some applications; however, 10 Mbps may even be too low for them. The problem is you can't reliably predict your application's network throughput.

Many AWS users have vociferously complained about inconsistent AWS network performance and many AWS competitors have criticized the company, citing their own network design and capability as superior to AWS's and thus providing users with a reason for switching services.

To be sure, Amazon should upgrade the infrastructure to provide higher, more reliable network capacity. However, one drawback is that it would impose higher costs on all AWS users, including the masses of users who have no concerns about typical AWS networking performance.

Consequently, Amazon has developed an AWS-like solution, rather than reconstructing the networking infrastructure from the ground up: an additional set of offerings to address the needs of applications requiring high-performance networking.

Management tools

Governance is associated with the directorial level of decisions while the management is the term that deals with the executive and managerial level of decisions within the organization, which here means AWS.

Amazon CloudWatch

Amazon CloudWatch is a kind of observing and management service developed for the site reliability engineers (SRE), system operators and the developers and IT managers. The CloudWatch caters you with the data, the actionable insights for keeping track of each activity of your application, respond to system-wide accomplishment changes and understand them, optimize the resource utilization, as well as achieve a unified representation of the Operational health. The CloudWatch captures the monitoring, operational data in the form of metrics, logs and events.

You can use the AWS resources, services and applications which run on the AWS and also on-premises servers. Also, you can use the CloudWatch for setting the high-resolution alarms, do the visualization of the logs as well metrics side by side., troubleshoot issues, take the automated actions, explore various insights for optimizing the applications, and make certain that they are running without any resolutions. Have a look below:

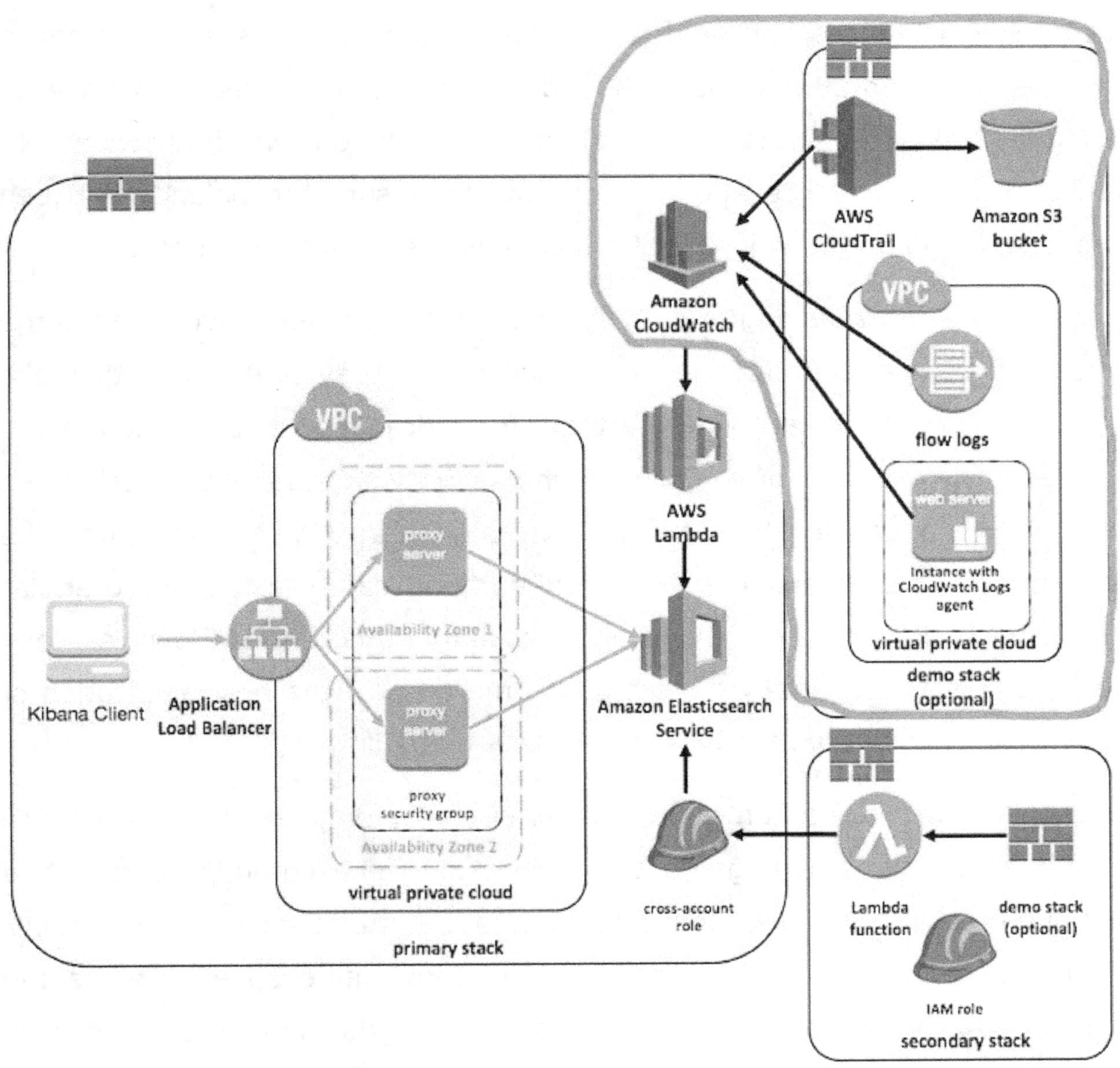

AWS Auto Scaling

The AWS Auto Scaling keeps a check on the applications and automatically tune the power for maintaining the steady, predictable accomplishment at the least possible cost. With the help of the AWS Auto Scaling, it's quite simple to set up the application scaling for the numerous resources over the numerous services in just a few minutes. The services cater a simple but powerful user interface which helps you develop the scaling plans for the resources which cover the

Amazon EC2 instances and spot fleets, the Amazon EC2 tasks, the Amazon Aurora and the Amazon DynamoDB indexes and tables. The Amazon Auto Scaling can make the scaling easy with recommendations which allows you to optimize the performance and the cost and to ensure the balance between them, also if you are making use of the compliances and the security.

And as the enterprise is migrated to the AWS, they normally have a very large number of applications and the distributed teams they often need to make numerous accounts for allowing the team to work independently, though still left is a consistent level of compliance and security. Besides, they use AWS's management in addition to security services such as AWS Architecture. The AWS Service Catalog and the AWS Configuration caters extremely granular controls for their workloads. They want to maintain this control. However, they also never want to centrally govern and come up with the best application of the AWS services via the accounts in the environments.

The central tower can automate the setup of the landing zones and configure the AWS security and management services that are based on the established awesome practices within a complaint, multi-account and secure environment. The distributed team can provision new accounts quite speedily while the central group members have the peace of mind as the new accounts can be centrally created. The company wide compliance policies give full control over the environment and without any form of sacrifice regarding speed or agility which the AWS issues over the environment, without compromising the agility or the speed which the AWS can issue your team meant for development. Have a look below:

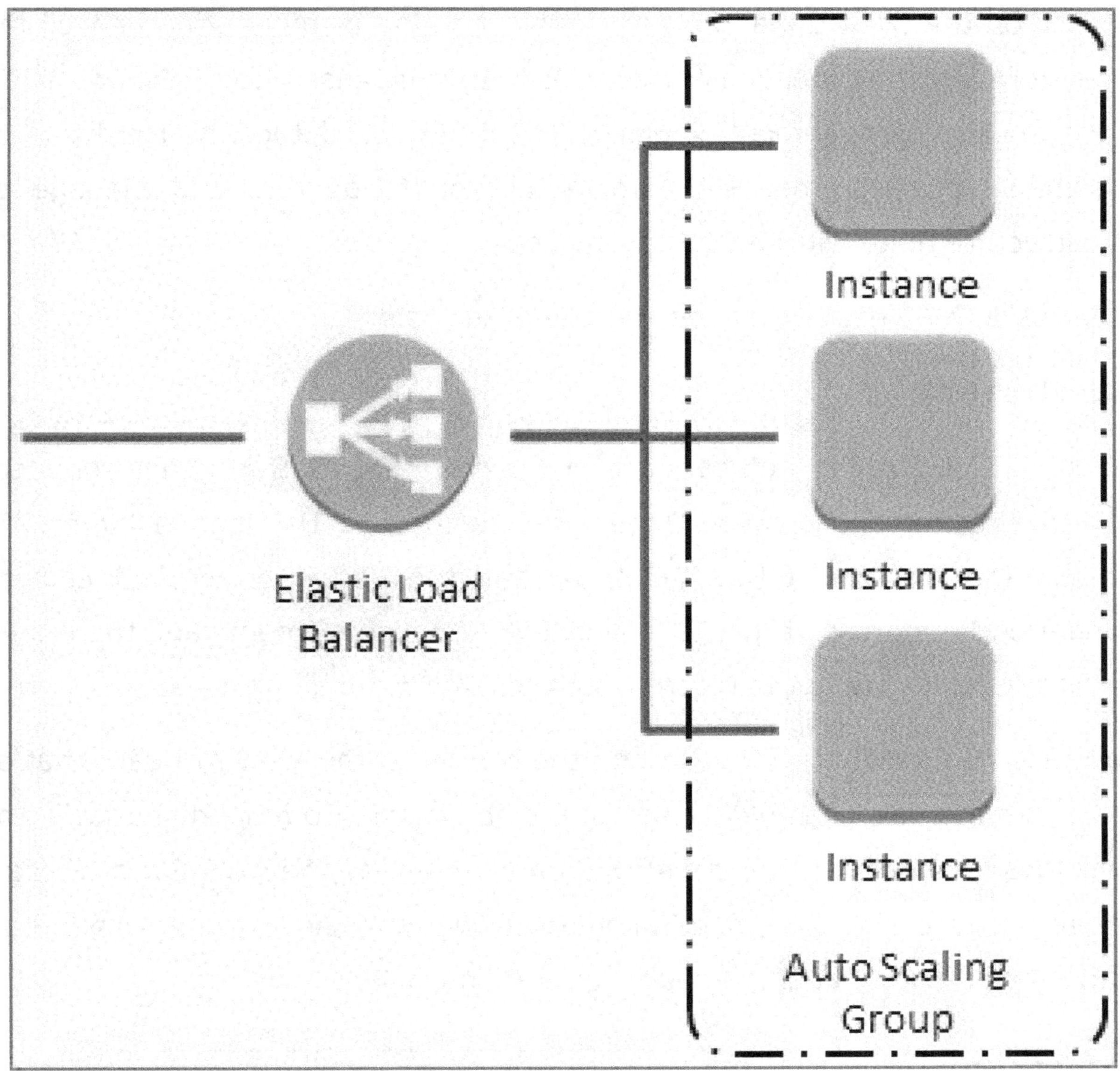

AWS Systems Manager

This can provide you the visibility and the control of the overall infrastructure which is established on the AWS, the system manager can unify the user interface and, hence, you can check the operational data from vary and any number of services and it permits you in automating the operational tasks over the AWS resources. You can join the resources like the Amazon S3 buckets, Amazon EC2 instances or the Amazon RDS instances, via the application, have a check on the operational data or troubleshooting or monitoring as well take the action over the collection of the resources. It makes it easier the application and resource management, reduces the time for detecting, resolving the operational puzzles, makes it simple for you to operate and manage the infrastructure quite with safety, at any scale.

AWS Systems Manager contains the following tools:

AWS CloudTrail

The AWS CloudTrail is web services that archive the AWS API calls connected to your account and transport the log files to you. The logged information shelters the identity of the API caller and the period of the API call, the foundation IP address of the API caller, the response elements and the request parameter which we get in return from the AWS Trail.

Via the AWS CloudTrail, you can find the history of the AWS API calls that are made with the help of the AWS management console, Command line tools, top-level AWS services such as the AWS CloudFormation. The AWS CLI call history is then produced by the CloudTrail which allows the security examination and compliance auditing or the resource alteration tracking.

AWS Config

The AWS Config is a fully supervised service which provides the AWS resource inventory, configuration history and the configuration change notification for enabling the governance and security, the configuration rules allows you to develop the rules which automatically tests the configuration of the AWS resources which are being recorded with the help of the AWS Config,

And with the AWS Config, you can find out the existing the AWS resources that have been deleted, which determine the overall compliance versus the rules, as well as dive into the configuration explanatory of the resources at any given instance of the time. These capabilities help you in enabling the compliance auditing, tracking of the resource changes, security analysis and troubleshooting.

AWS OpsWorks

The AWS OpsWorks is a configuration supervising service that caters to the managed instances of the Chef and the Puppet. These allow the automation of the platforms which leverages you to automate the configuration through coding, The OpsWorks allows you to use of the Chef and Bucket automate the server configuration, deployment, the supervision over the Amazon EC2 instances or like on-premises, compute atmospheres. The OpsWork happens to have the three offerings like AWS OpsWorks for chef automation, AWS OpsWorks Stacks and the AWS OpsWork for Puppet Enterprises.

AWS Service Catalog

The AWS Service Catalog helps the organization for creating and managing the catalogs for the IT services which are being approved for the application on the AWS. These IT services can cover almost everything from the VM images, software, servers, databases to overall multi-tier application architecture. The AWS allows consistent governance and can meet your compliance necessities, while you enable your users for quickly deploying the approved IT services that they require,

AWS Trusted Advisor

The AWS Trusted Advisor is also an online resource that caters you to reduce cost, make better performance, better security for optimizing the AWS environment; the trusted Advisor can leverage the real-time guidance for helping you to provision your resources via the AWS best ethics.

AWS Personal Health Dashboard

This one caters you alerts and remediation guidance while the AWS experiences the events which can affect you, while the service health Dashboard shows the general state of the AWS services, the Personal Health Dashboard allows you to personalize the view to the performance and the availability of the AWS services which underlies your AWS resource. The dashboard also displays you the related and the time-based information for catering you the management skills to for managing the events that are in progress and this caters to the proactive notification to help you with the plan of the activities scheduled with the personal health dashboard, the alerts are without any human intervention triggered and the changes made in the health of the AWS resources can give you the visual of the event, guidance diagnose and also resolving the issues.

AWS Managed Services

This provides you the going on the management of the AWS infrastructure such that you can focus on the application. Through the implementation of the best ethics for maintaining your infrastructure, the AWS managed services can cater to drop short the operational overhead and the risk. The AWS managed services also automate the various common activities like the change requests, monitoring the patch management, backup services, control the help for enforcing the corporate, security infrastructure policies and enables you to develop the solutions and the applications with help of the preferred development approach. The AWS supervised services also improve the agility, unburdens you from various infrastructure operations and reduce the cost, such that you can direct the resources towards making your business better.

AWS Console Mobile Application

The AWS Console mobile application can let your customer view and manage the select set of resources for supporting the incident responses while you are on the go. The console mobile application caters AWS consumers to monitor the resources via the dedicated dashboard, view the configuration details, alarms for select AWS services and the metrics. The dashboard provides permission to the users with resource status and real-time data on the Amazon CloudWatch, AWS Billing, cost management, Personal health dashboard with a single view. The customers can also view the ongoing issues, follow via the relevant CloudWatch the alarm screen for a detailed view through graphs and options for configurations. In addition, you can check the state of your specific AWS services, a view the resource screens with all the details, as do the select actions.

AWS License Manager

This makes your task easier to manage the licenses in the AWS and on-premises servers from software vendors like SAP, IBM, Microsoft and the Oracle. The AWS License Manager allows the administrators to develop the customized licensing rules which emulate the terms of the licensing agreements and then implementing these rules when an instance of the EC2 gets launched. The administrators can use of these sets of rules for limiting the licenses for different servers on a short-term basis. The rules in the AWS License Manager can enable you to put a limit on the violations in the licensing like using more of them than what agreed and reassigning the licenses to various servers on the short-term basis. The rules related to the AWS Licensing can help you to limit the licensing breach through stopping the instance physically from launching or by notifying the administrators regarding the infringement, administrator gain control and visibility of the licenses with the help of the AWS license Manager dashboard besides reducing the amount of risk of the misreporting, non-compliance, the additional cost related to the licensing overages,

The License Manager joins with the AWS services for easing the management of the licenses over the numerous AWS accounts, IT Catalogue and on-premises, via the single AWS accounts. The license administrators can also add various rules in the AWS Service Catalog, that allows you to create and manage the IT services which are approved for the applications on all the AWS accounts via the seamless integrations together with the AWS system manager and AWS Organizations, the administrators can manage the licenses over all the AWS accounts in the organization and on-premises atmosphere, The AWS marketplace buyers can now use of the License Managers for tracking the BYOL software which is obtained from the marketplace and keep the consolidated view of each of the licenses,

AWS Well-Architected Tool

This helps to review the current state of the workload, compare them with the latest AWS architecture most awesome practices, the tool bases on the AWS Well Architected Framework, which is developed for helping the cloud architects to develop the more secure, high in performance, resilient and good in efficiency the application infrastructure. This framework caters the consistent approach for the customers as well as the partners for evaluating the architecture and has been applied in hundreds of hundreds of workload reviews which has been conducted by the AWS architecture team, it also caters the guidance to help out in implementing the designs for scaling with the application requirements over the time.

AWS Security, Identity & Compliance

Security and Compliance Services

AWS provides various organizations with security and compliance services for uninterrupted services. Security and compliance services can be linked with other AWS services as well. The cloud computing security service is a very rapid growing service that functions similarly to the traditional style of IT security. This includes the protection of all the critical information from data leakage, deletion, and theft.

Security

Security is regarded as the highest priority service of AWS. As a user of AWS services, you will be gaining profit from a data center and a network architecture, which has been built to meet all the requirements of the highest organizations, which are security sensitive. While using the cloud, securing it from all possible sides is of utter importance. Cloud security is somewhat like the security of your on-premises data center. It is of much more important than the cost of the hardware and maintenance facilities. You are not required to manage the storage devices or physical servers in the cloud. Instead of doing it yourself, you use security tools that are based on software for protecting and monitoring the overall flow of information that goes in and out of the cloud resources.

One of the most interesting and advantageous features of AWS Cloud is that it allows the users to scale along with innovating. It can be done while maintaining a super-secure environment, without paying for those services that you do not use at all. This means that you can enjoy premium security and that too at a lesser cost when compared to the security of your on-premises data center environment. While using the services from AWS, you need to inherit all the practices regarding AWS policies, operational processes, and architecture which have been built for satisfying the ultimate requirements of those customers who are the most security sensitive. You can enjoy the agility along with flexibility with security services from AWS for your data centers.

AWS Cloud comes with a shared responsibility model. While the security of the cloud is managed by AWS, you are the one who is completely responsible for the security in the cloud. In simple words, you can retain the security control which you choose for protecting your content, applications, platform, networks, and systems in the same way that you would have done for an on-premises data center. You will also receive the required guidance and competence via online personnel, resources, and partners. AWS provides users with various advisories for current issues. You can also enjoy the opportunity to work along with AWS whenever you come across any kind of security issue.

For the purpose of meeting up with your objectives regarding cloud security, you also get access to various features and tools. AWS provides the users with tools and features specific to security across network configuration, access control, management and encryption of data. The AWS environments are audited at regular intervals with certifications from various accreditation bodies across verticals and geographies. You can take full advantage of the automated tools in the AWS environment for the purpose of access reporting and asset inventory.

Benefits of AWS security

AWS security comes with various benefits that can make your cloud working much more convenient and secure.

It helps in keeping all your data safe. The infrastructure of AWS puts in place strong safeguards for protecting the privacy of the users. All of your data is stored in super-secure data centers of AWS.

You can meet compliance requirements with AWS security. AWS is known for managing various programs of compliance within its infrastructure. In simple words, the segments of your required compliance have been completed already.

You can save lots of money with AWS security services. You can cut down your costs with the help of AWS data centers. You can maintain the highest security standards without any need of managing your facility.

You can scale quickly and conveniently with AWS security services. The security services also scale with the usage of your AWS Cloud. Whatever may be the size of your business, the infrastructure of AWS has been designed for keeping all your data safe.

Infrastructure security

AWS provides users with various capabilities in security and services for improving the control and privacy of network access. These also include:

Customer-controlled encryption, which is in transit with all the TLS across all the services of AWS.

The network firewalls which are built into VPC from Amazon and firewall capabilities of the web applications in AWS WAF lets you create various private networks and also control access to the applications and instances.

Options of connectivity that enables dedicated or private connections from your dedicated on-premises environment or office.

Automated encryption of all your application traffic on the regional and AWS global networks between the secured AWS facilities.

Encryption of data

With the AWS security service, you can get the facility of adding an extra security layer to all the data at rest inside your cloud along with providing efficient encryption and scalable features. This also includes:

Flexible options of key management along with AWS KMS or key management service. This service will allow you to choose if you want AWS to manage all the encryption keys or you want to have complete control over the keys.

Capabilities of data encryption available in database and AWS storage services such as S3, EBS, Glacier, SQL RDS, Redshift and Oracle RDS.

Queues for encrypted messages required for the transmission of highly sensitive data using SSE or server-side encryption for SQS.

Dedicated cryptographic key storage based on hardware using CloudHSM that will allow you to satisfy all your compliance needs.

Additionally, AWS also provides APIs for integrating data protection and encryption with any kind of service that you deploy or develop in the AWS environment.

Configuration and inventory

AWS offers the users with a wide range of useful tools that will allow you faster movement while still safeguarding the resources of your cloud with the best practices and organizational standards. This service includes:

Various tools for deployment and for managing the decommissioning and creation of the AWS resources according to the standards of the organizations.

Amazon Inspector, which is a security assessment service from AWS that assesses all the applications automatically for deviations or susceptibility from the best practices that also include OS, impacted networks and attached storage.

Tools for configuration and inventory management along with AWS Config that will identify the resources of AWS and then manage and track all the changes to all those resources with time.

Tools for template management and definition along with AWS CloudFormation for creating preconfigured and standard environments.

Logging and monitoring

AWS security provides you with various features and tools with which you can have a complete visualization of what is happening in the environment of AWS. This also includes:

Options for log aggregation, compliance reporting and streamlining the investigations.

Usage of AWS CloudTrail for having deep visibility into the API calls that also includes who, when, what and from where all the calls were made.

Alert notifications with the help of Amazon CloudWatch whenever thresholds exceed, or any specific event occurs.

All these tools will provide you with the ultimate visibility that you need for spotting any kind of issues right before they can impact the functioning of the business. It also allows the users to improve the posture of security along with reducing the risk of the environment.

Access and identity control

AWS security services offer you all the capabilities for defining, managing and enforcing the policies of user access across all the services of AWS. This also includes:

AWS IAM or Identity and Access Management for defining the respective user account with proper permissions across the resources of AWS.

AWS Directory Service that allows you to federate and integrate with the various corporate directories for improving the experience of end-user and reducing the overhead of administration.

AWS Multi-Factor Authentication for all the privileged accounts along with options for the authenticators, which are based on hardware.

AWS provides the customers with native access and identity management integration across most of its services that also includes the integration of API with any of your services or applications.

Penetration testing

AWS regularly tests its own infrastructure, the results of which can be found in the compliance reports. The customers of AWS security services that easily carry out assessments of security and penetration testing against the AWS infrastructure of their own without any kind of prior approval for the core number of services.

DDoS Mitigation

Availability is of preeminent importance for the cloud. The customers of AWS benefits from the services of AWS and technologies, which has been built for providing flexibility during the time of DDoS attacks. With a combination of the AWS security services, it is possible to implement a well-built defense with an in-depth strategy for countering the DDoS attacks. The services, which have been built for DDoS help, help in minimizing the overall time of mitigation and also reduces the impact.

AWS Compliance

With the help of AWS compliance, you can easily understand the various robust controls, which are in place at AWS for maintaining the overall data protection and security in the AWS cloud. As the systems or applications are, built on top of the infrastructure of AWS Cloud, the responsibilities of compliance need to be and will be shared. By tying altogether, the audit-friendly and governance-focused service features along with applicable audit or standards of compliance, the enablers of AWS Compliance builds upon the traditional programs. This, in turn, helps all the customers to operate and establish in an environment that is controlled by AWS security.

The IT infrastructure provided by AWS to its customers is managed and designed in alignment with the best practices of security along with a variety of standards in IT security. The following are the partial assurance programs, which are satisfied by AWS:

FISMA, FedRAMP, and DIACAP

SOC 2, SOC 1/ISAE 3402 and SOC 3

ISO 9001, ISO 27017, ISO 27001 and ISO 27018

PCI DSS Level 1

AWS provides its customers with a wide variety of information on the IT control environment in reports, whitepapers, accreditations, certifications and other types of third-party attestations.

Attestations and Certifications

The certifications of attestations and compliance are determined by an independent, third-party auditor, which results in an audit report, certification or compliance attestation. The auditors include:

C5

ASIP HDS

DoD SRG

FIPS

ISO 9001

ISO 27017

ISO 27001

ISO 27018

TISAX

K-ISMS

PCI DSS Level 1

Privacy and regulations

The customers of AWS remain responsible for adhering to the compliance regulations and laws. In some of the cases, AWS also offers enablers, functionality and legal agreements such as AWS Business Associate Addendum for supporting the compliance of the customers. No kind of formal certification is feasible to the providers of cloud service within these regulatory and law domains.

CCPA

CLOUD Act

CISPE

FERPA

GLBA

HIPAA

GDPR

ITAR

VPAT / Section 508

PIPEDA

PHIA

PDPA – 2010

Privacy Act (New Zealand)

IRS 1075

Frameworks and alignments

The frameworks and alignments of compliance include compliance or published security requirements for a particular purpose such as a particular function or industry. AWS provides users with enablers and functionality for these kinds of programs. The requirements under particular frameworks and alignments might not be subject to attestation or certification. However, some of the frameworks and alignments are covered under other programs of compliance.

CJIS

CIS

CSA

FFIEC

EU-US Privacy Shield

FISMA

ICREA

MPAA

NIST

MITA 3.0

UK Cloud Security Principles

G – Cloud

Uptime Institute Tiers

Application Integration

Amazon SQS

SQS or Simple Queue Service is designed to manage message queuing services. Simple Queue Service is capable of sending messages across platforms and multiple services. These services that can have messages sent to them include EC2 instances, S3, and DynamoDB. It makes moving messages from application to application possible, no matter the state of the service (whether it is active or inactive). The SQS service has a visibility timeout of up to 12 hours. The service delivers information by making use of the Java message queue service.

Amazon Web Services Step Functions

The Step functions tool allows users to have a convenient method of coordinating the various components of applications and services using a visual workflow portrayal. This tool helps by coordinating components and steps through the various functions of the client's application, by creating a graphics console for the user to be able to arrange and visualize the separate components as a multitude of steps, making it easier to build and run more complex applications. In addition, the Step functions service automatically tracks each step as it is triggered, automatically retrying when errors occur, allowing the application to be executed in order. The service logs each step, allowing for easier diagnosis and debugging, and steps can be changed or added or removed without the need for additional coding.

Simple Notification Service or SNS

SNS is a great service to work with when it comes to the AWS system. This is going to be a service that is able to send out notifications about an event through a convenient mechanism as a way to alert a person or a program that something of interest has just happened. The simplicity of this description is going to show us the power of notifications, however. Think about the administrator of a system who is going to need to make sure that any application in AWS is working in the proper manner. If there is something that is no longer working, then this person needs to know about it right away.

One way to make the administrator figure out about the problems in an application would be to have them be always logged into AWS and to check over the state of the application all of the time. Of course, this is not really practical, and no one wants to spend that time with the application. So, we can work with another option that will allow us to define one or more conditions that the administrator has to know about and then create a mechanism that is going to respond to these conditions.

When the machine does what it is supposed to, it is going to look around for one of these conditions, such as an error that shows up in a reading of the database. If this is found, then the mechanisms going to send out a notification to one or more people so that they are able to determine whether this is worth their time to take some action.

These notifications can be sent through a lot of different methods based on what works for you. And in some cases, we do not have to send out these notifications to a human because we are able to send them out to a program that is capable of handling this issue and taking care of the message as it needs to.

Notifications are going to be a really simple idea to work with, but they are powerful. Many system administrators like to work with these and like how simple it is to stay on top of the system without having to stare at it all of the time. This is what SNS is going to be able to help us out with along the way.

These are just a few of the services that we are going to be able to enjoy when it comes to working with the AWS services for our needs. Amazon has really created a great product that allows us to have some freedom in the work that we are doing and will ensure that we are able to reach out and make our business as strong as possible.

There may be other similar services out there that we are able to work with, but none are going to provide us so many services in one place, and none are going to ensure that we are set up and ready to go for an affordable price. Those other options are not able to stand up to the Amazon option, and you can use AWS, even if you do not need all of the services, and see some of the amazing things that are possible with Amazon.

Amazon SQS

SQS or Simple Queue Service is designed to manage message queuing services. Simple Queue Service is capable of sending messages across platforms and multiple services. These services that can have messages sent to them include EC2 instances, S3, and DynamoDB. It makes moving messages from application to application possible, no matter the state of the service (whether it is active or inactive). The SQS service has a visibility timeout of up to 12 hours. The service delivers information by making use of the Java message queue service.

Amazon API Gateway

Amazon's API gateway is a service designed to make it much easier for Amazon clients to develop, publish, monitor, maintain, and keep secure their APIs, regardless of the scale. This tool can be accessed through the Amazon Web Services management console, by creating an API that acts as a door for applications to be able to access data, logic, or other functionalities from the client's chosen back – end services such as EC2 or lambda, or even web applications. The API gateway function automates all the tasks needed in order to process hundreds of thousands of API calls, including but not limited to monitoring, authorization, access control, and traffic management.

AWS Desktop & App Streaming

Amazon WorkSpaces

This service can easily be accessed through the Amazon Web Services management console, allowing for quick and easy deployment of cloud desktops for a multitude of users.

Amazon WorkSpaces is an overseen, secure Desktop-as-a-Service (DaaS) arrangement. You can utilize Amazon WorkSpaces to arrangement either Windows or Linux work areas in only a couple of moments and rapidly scale to give thousands of work areas to laborers over the globe. You can pay either month to month or hourly, only for the WorkSpaces you dispatch, which encourages you to set aside cash when contrasted with conventional work areas and on-premises VDI arrangements. Amazon WorkSpaces causes you to dispose of the intricacy in overseeing equipment stock, OS forms and fixes, and Virtual Desktop Infrastructure (VDI), which improves your work area conveyance methodology. With Amazon WorkSpaces, your clients get a quick, responsive work area of their decision that they can get to any place, whenever, from any upheld gadget.

It is possible to utilize this service by paying for it hourly or monthly, just for the WorkSpaces that you decide to launch, which helps you to save money when compared to some of the traditional options that you have for desktops. It is one of the best options to work with because it is going to eliminate some of the complexity that comes with managing your hardware inventory, OS versions, and patches, and the VDI, which is going to make the delivery strategy that you do with your desktop that much easier. Plus, it is fast and more responsive compared to other options, so we are able to get the best results with it in no time.

Amazon AppStream 2.0

The Appstream 2.0 is a streaming service built by amazon that allows the streaming of any desktop application on the Amazon Web Services cloud to any device that is capable of running a web browser, without need for any rewriting. The service allows for instant access to the needed applications, regardless of the device being used. While we know that compatibility issues for a lot of applications prevent smooth usage or transition from one device to another, the AppStream 2.0 service of amazon provides a solution by letting users enjoy the benefits of a native browser application without need for re – writing the application's code, provided that the device can run an HTML – 5 compatible browser. This allows developers to simply maintain one version of their applications, allowing users to access the latest version, while keeping application maintenance and patching issues to a minimum, and reducing the stress of having to deal with porting the application for compatibility to different platforms. Furthermore, the AppStream 2.0 allows for instant and global scaling on a pay – as - you –use model, making it convenient for developers to use.

Amazon App Stream 2.0 is a completely overseen application gushing service. You halfway deal with your work area applications on App Stream 2.0 and safely convey them to any computer. You can undoubtedly scale to any number of clients over the globe without procuring, provisioning, and working equipment or infrastructure. App Stream 2.0 is based on AWS, so you profit by a server farm and system engineering intended for the most security-delicate associations. Every client has a liquid and responsive involvement in your applications, including GPU-concentrated 3D structure and designing ones, because your applications run on virtual machines (VMs) upgraded for explicit use cases and each spilling session consequently changes with arranging conditions.

Endeavors can utilize App Stream 2.0 to streamline application conveyance and complete their migration to the cloud. Instructive establishments can give each understudy access to the applications they require for class on any computer. Programming merchants can utilize App Stream 2.0 to convey preliminaries, demos, and preparing for their applications with no downloads or establishments. They can likewise build up a full software-as-a-service (SaaS) arrangement without modifying their application.

Amazon WorkLink

The Amazon WorkLink is a completely supervised service which can let you cater to your employees with safety, easy to access your internal business website and apps with the help of their mobile phones; Traditional solutions like Virtual Private Networks (VPNs) or device management software are troublesome for use on the go, again and again, need the custom browser which shows poor user experience. As a result, the employees also detriment with the help of often relinquish consuming them overall.

With the Amazon WorkLink employees can access to internal web content as effortlessly as they get the accessibility of any website without problems like connecting with their commercial network. When the user uses the internal website, the webpage is at first rendered through a browser that runs on the protected container in the AWS. The Amazon WorkLink then transfers the content of this page to the phone employees in the form of vector graphics meanwhile keeping safe the functionality and the interactivity of the page. This approach is safer than the old-style solutions as the internal content is not ever stored or being cached via the browser on the employee phones, besides the employee devices can never get connected straightforwardly to the business network.

And with the Amazon WorkLink, you need not require the minimum fees on the very long-term promises: you will never find any additional charges for consuming the bandwidth.

Developer Tools

AWS Code Commit

Code Commit is AWS's version control administration that enables you to store your code and different resources secretly in the cloud. It is a completely overseen source control administration that hosts secure Git-based storehouses. It makes it simple for teams to work together on code in a safe and exceptionally versatile environment. Code Commit dispenses with the need to work your very own source control system or stress over scaling its infrastructure. You can utilize Code Commit to safely store anything from source code to pairs, and it works flawlessly with your current Git devices.

AWS Code Build

AWS Code Build mechanizes the way toward building (compiling) your code. It is a completely overseen assemble administration that compiles source code, runs tests, and produces software bundles that are prepared to deploy. With Code Build, you don't have to provide, scale, and maintain your own build servers. This administration scales consistently and forms numerous builds simultaneously, so your builds are not left holding up in a line. You can begin rapidly by utilizing prepackaged form situations, or you can make custom build conditions that utilization your own build apparatuses.

AWS Code Deploy

This is a method for sending your code on EC2 instances naturally. It is a service that can automate code deployment, including instances running on-premises and EC2 instances. Code Deploy makes it simpler for you to quickly release new features, encourages you to stay away from personal time during application deployment, and handles the toughness of updating your applications. You can utilize Code Deploy to robotize programming arrangements, vanishing the requirement for failure inclined manual activities. The administration scales with your infrastructure so you can, undoubtedly, convey to one instance or thousands.

AWS Code Pipeline

Code Pipeline enables you to monitor various strides in your organization like structure, testing, confirmation, and arrangement on advancement and Production conditions. Code Pipeline is a completely overseen nonstop conveyance service that causes you to mechanize your release pipelines for quick and dependable application and framework updates. It computerizes the build, test, and send periods of your release procedure each time there is a code change, in view of the release model you characterize. This empowers you to quickly and dependably convey features and updates. You can, undoubtedly, incorporate Code Pipeline with third-party administrations, for example, GitHub or with your own custom module. With AWS Code Pipeline, you pay for what you use. There are no forthright expenses or long-term responsibilities.

AWS Code Star

AWS Code Star is a cloud-based aid for making, overseeing, and working with software advancement projects on AWS. You can rapidly create, manufacture, and convey applications on AWS with an AWS Code Star venture. Code Star empowers you to quickly create, construct, and send applications on AWS. It also gives a unified user interface UI, empowering you to effortlessly deal with your product improvement exercises in a single spot. With Code Star, you can set up your continuous conveyance toolchain in minutes, enabling you to begin to release code quicker. In addition, Code Star makes it simple for your entire group to cooperate safely, enabling you to effectively oversee access and add owners, and viewers to your tasks. Each Code Star venture accompanies a project management dashboard, including a coordinated issue following capacity powered by Atlassian JIRA Software. With the Code Star venture dashboard, you can track progress over your entire software development process, from your excess of work items to groups' ongoing code deployment.

Amazon Corretto

Corretto is a no-cost, multiplatform, creation-ready distribution of the Open Java Development Kit (Open JDK). Corretto accompanies long term support that will incorporate execution upgrades and security fixes. Amazon runs Corretto inside production administrations and it is guaranteed as good with the Java SE standard. With Corretto, you can create and run Java applications on famous operating systems, including Linux 2, Windows, and macOS.

AWS X-Ray

AWS X-Ray makes it simple for engineers to break down the conduct of their disseminated applications by giving solicitation following, special case assortment, and profiling capabilities. This service assists engineers with analyzing and troubleshooting dispersed applications in production or a work-in-progress, for example, those constructed utilizing a microservices design. With X-Ray, you can see how your application and its basic administrations are performing so you can recognize and investigate the root cause of execution issues and blunders. X-Ray gives an end-to-end perspective on demands as they travel through your application and shows a guide to your application's hidden parts. You can utilize X-Ray to dissect the two applications being developed and in production, from basic three-level applications to complex microservices applications comprising thousands of administrations.

Introduction to Amazon CloudWatch

- What is Amazon Cloud Watch?
- How does Amazon Cloud Watch work?
- Components of CloudWatch

Amazon Cloudwatch is one of the commonly used services provided by Amazon that keeps a check on AWS applications and resources on the AWS cloud.

What is Amazon Cloud Watch?

Amazon Cloud Watch is a management service designed to monitor the resources and applications you use by collecting metrics/logs which are variables that defines the performance measure.

On the Cloudwatch homepage, the metrics are displayed for every AWS service you use. You can also create a custom dashboard that shows metrics of your custom applications. This service is beneficial for developers, IT managers, and engineers as it makes it easier to do diagnostics and troubleshooting to ensure applications are working seamlessly.

You have the option to create alarms that notifies you whenever there are changes made to the resources you're monitoring when a threshold is compromised. For instance, you can check the CPU usage status, and using this information will help you in deciding whether you should load additional resources or not. You can use 10 custom metrics and alarms for free. You'll be charged for additional metrics.

How does Amazon Cloud watch work?

Amazon Cloudwatch follows a sequence of actions - first, it collects logs and data, then it closely monitors the applications in realtime and acts according to the rules and then, in the end, it analyzes the metrics for further use. During the collection of data, Amazon Cloudwatch continuously monitors and updates the logs. If in case any issues are found, troubleshooting is started.

To put it simply, the service is like a repository of metrics and you can retrieve statistics and visual analysis based on those metrics.

The computing of resources is done in powerful and huge data centers. To ensure flexibility and scalability, each and every data center is located in different regions and each region is separate from other regions so as to achieve stability and prevent system failure. Metrics are also stored in each region so you could analyze the statistics of working in each region.

Components of CloudWatch

Namespaces

A namespace is like a container to hold metrics. Metrics found in different namespaces are treated separately so that there is no overlapping.

When you're creating a metric, you specify a namespace that must have valid characters (alphanumeric, underscore, slash, hash, etc.) The AWS namespace follows a typical name convention such as AWS/example.

Metrics

Metric is the major component in CloudWatch. It corresponds to a variable that is monitored, and the values of a variable represent the data points.

You can view the built-in metrics provided by AWS or you can send your custom-made metrics to cloud watch. Metrics only exists in the region in which they are made. They can be deleted; however, they expire after 15 months if there are no data points. You can define a metric by name or namespace.

Time Stamp

Timestamp refers to date and time that is associated with a metric data point. If you don't create a time stamp, CloudWatch automatically sets up the time stamp based on when the data point was received. The timestamp shows the complete date, hours, minutes and seconds.

Dimension

A dimension refers to the characteristics that describe a metric. Dimensions make it easier to locate a certain metric and obtain its statistics. For instance, if you want to get statistics for a particular EC2 instance, you can specify the dimension for the instanceID.

There are some metrics generated by other AWS services such as Amazon EC2 for which CloudWatch can aggregate data and display the statistics. If you're searching for metrics in the AWS/EC2 namespace but don't mention any dimensions, CloudWatch will aggregate all of the data for the mentioned metric and output the statistics.

Statistics

Cloudwatch provides data aggregation over a specific period. Data aggregation is the collection of related things altogether and it's computed by using the metric name, dimension, namespace, and data point.

AWS provides a list of available statistics as mentioned below:

Minimum

It's the lowest value that is observed during the specified period.

Maximum

It's the maximum or highest value that is observed during the specified period.

Sum

Values belonging to the same metric are added together. This statistic helps determine the total volume of a metric.

Average

It's the ratio of sum and sample count during the specified period. By comparing the value of average to Minimum and Maximum, it will help you in deciding when to increase or decrease the resources.

Sample count

It records the number of data points.

pNN.NN

It's the value of the specified percentile. Using up to 2 decimal places, you can type any percentile.

Units

Each statistic is measured in units which could be seconds, percent, bytes, etc. Whenever you create a custom-made metric, it's a good practice to specify the unit. If you don't mention any unit, Cloud watch uses none as the unit. By specifying units, you make the data more meaningful. Cloud watch will aggregate similar data having the same units and output the results. If you don't specify a unit, Cloud watch will treat all the data same.

Period

A period refers to the time duration associated with a statistic. Each statistic shows data aggregation over a specific time frame. It's measured in seconds and the valid values for period are multiples of 60.

If you want to retrieve statistics of any metric, you can mention parameters like start time, end time and period. These variables return the overall length of time of the corresponding statistics. The default values of the parameters will output a collective set of statistics of the previous hour.

When you aggregate the statistics over a time frame, they are marked with time corresponding to the starting time. For example, data aggregated from 10:00 pm-11:00pm is marked as 10:00pm. Furthermore, data aggregated between 10:00 pm – 11:00pm becomes visible at 10:00pm. The aggregated data may change as more samples are being collected during the time duration.

Periods can be used for alarms. When you want to monitor a metric, the cloud watch compares the metric to the threshold value that a user specifies. You can specify the time over which the comparison takes place, and you can mention how many evaluation periods are needed to reach to the conclusion.

Aggregation

Statistics can be aggregated on the basis of period you specify when you're retrieving statistics. You're allowed to add data points having similar time period so that AWS CloudWatch can aggregate them. CloudWatch doesn't aggregate data according to the region.

Not only you can aggregate data according to the period, but you can do the same for dimensions and namespace. You can add data points for same or different metrics, with different time frames.

In a distributed system, metrics from different sources having same dimensions and namespaces are treated as a single metric.

Percentile

A percentile tells the position of a value in a dataset. Now you might be wondering why and where percentiles are used? Whenever you're running a website or an application, you need to make sure if it's running to its maximum potential to provide a greater user experience. When you're looking at the averages, you may not be able to see the big picture because the average may not include the outliers, for example, that 5% of the users didn't have a good experience.

Percentiles are indeed a useful statistic that helps you to understand the behavior and performance of the application you're running. Some CloudWatch metrics support the use of percentiles. You can specify the percentile up to 2 decimal places and use it when creating an alarm.

Percentile is available for custom metric as well provided that you publish the un-summarized data point. However, you can't use it for metric values, which are negative.

Alarm

An alarm is used to invoke actions automatically. It can watch a single metric over a specific period and performs an action based on the metric value against the threshold value.

The action is a notification that is sent to Amazon SNS or Auto scaling policy. There is an option to add an alarm on the dashboard.

When you're creating an alarm, select a period greater or equal to the frequency of the metric that is monitored. For example, detailed monitoring provides metrics for your EC2 instances every 1 minute. This means that you need to set an alarm of 1 minute (60 seconds).

Conclusion

Amazon is providing a lot of web services. Many of them are stated above.

Cloud computing, in the simplest terms, means storing and processing data and services over the Internet, rather than the hard drive of your computer.

Your hard disk is what cloud computing isn't about. That's called local storage and computing when you store data on-or run programs from the hard drive. Anything you need is physically close to you, which means you have fast and easy access to your data (for that one device, or those on the local network). Operating off the hard drive is how the tech industry has been operating for a decade.

The future of computing technology lies in the cloud. Which means that if you're not adapting your company to suit the cloud model your company will be left behind in this world of modern technology.

Cloud computing is when organizations share a network of freely accessible servers. Servers are stored on the Internet, allowing companies to handle data "in the cloud" instead of on a local server. It is a virtual space in which devices on the network can access data from anywhere.

While cloud computing has only picked up big momentum over the last two decades or so, the concept has been around since the 1960s. John McCarthy, a renowned computer scientist, introduced the idea when he invented a technology that would allow computation to be marketed as a commodity such as electricity or water. He indicated that each subscriber would only have to pay for the capacity they actually used and that certain users would be able to sell services to other users.

Amazon Web Services (AWS) Introduction to Famous Amazon Web Services is a robust cloud platform developed by Amazon's e-commerce giant. It offers software-as - a-Service (SaaS), platform-as - a-Service (PaaS), and infrastructure-as - a-Service (IaaS) services. Think about the history of the electricity supply to grasp the logic of AWS.

Initially, factories will build their own plants to fuel their own facilities. Over time, governments and private investors have developed large power plants that supply electricity to numerous towns, factories, and homes. Under this new model, the factories will pay even less per unit of power due to the economies of scale enjoyed by the massive power plants. AWS was designed and built on the basis of a similar logic.

By 2006, Amazon had established itself as the world's largest online retailer, a role it still holds. Seamlessly running such a huge operation required a large and sophisticated infrastructure. Its imbued Amazon with deep expertise in the management of large-scale network and server networks.

As a result, AWS was launched in 2006 as Amazon tried to make accessible to companies and individuals the technology infrastructure it had developed and the expertise it had gained. AWS was one of the first pay-as-you-go (PAYG) computing models that could scale performance, storage, and computing based on the evolving needs of the user.

Amazon Web Services offers cloud infrastructure from hundreds of data centers and numerous availability zones (AZs) spanning regions of the world. Every AZ includes a number of data centers. Customers can set up virtual machines and duplicate their data in several AZs in order to provide a highly scalable network that is resistant to server failure or data center failure.